Official Journal of the J.B. Johnston Club

Brain, Behavior and Evolution

KARGER

Printed in Switzerland
on acid-free paper by
Reinhardt Druck, Basel

Appears monthly:
2 volumes per year
(12 issues)

Guidelines for Authors

Brain, Behavior and Evolution

Brain, Behavior and Evolution publishes original research reports on the structure, function and evolution of the nervous system in vertebrates and invertebrates. The journal welcomes studies from the fields of comparative neuroanatomy, comparative neurophysiology, and neuroethology, as well as studies in behavioral biology and functional morphology that relate to the organization or evolution of the nervous system. Submitted papers need not be integrative or multidisciplinary, but they should contribute to a broader understanding of nervous system evolution by providing information that is relevant from a comparative or evolutionary perspective. The journal will also consider review papers, theoretical papers, and symposium proceedings; authors of such works must contact the editor prior to submission to determine if the papers are appropriate to the journal. In all cases, relevance and merit are the criteria for acceptance.

Submission

Only original papers written in English are considered. Standard American English spelling and grammar should be used. If English is not the first language of the author(s), it is strongly recommended that an English-speaking colleague be asked to go over the paper before it is submitted. Manuscripts should be sent to:

Prof. Walter Wilczynski
Department of Psychology
University of Texas
4.212 Seay Psychology Building
108 E. Dean Keaton
Austin, TX 78712 (USA)
wilczynski@psy.utexas.edu

Manuscripts should be submitted in quadruplicate, with 4 sets of illustrations of which at least 1 is an original or a facsimile sufficient to indicate reproductive quality. (Original artwork, particularly oversized illustrations, may be retained by the author until requested by the editor.) The entire manuscript should be typed on one side of the paper. *No part of the manuscript should be less than double-spaced with wide margins.* American letter-size ($8^1/_2 \times 11$ in.) paper is preferable; if this is not available, authors using European A4-size (210×297 mm) should set margins accordingly (i.e. at least 1 inch at the top and sides, 2 inch at the bottom). Do not divide a word at the end of a line. Do not justify the right margin of the text; it is far more important to maintain accurate spacing and avoid misleading hyphenations. Manuscripts that are single-spaced, poorly reproduced, or otherwise difficult to read will be returned, unreviewed, for repreparation. Authors are strongly encouraged to include a disk version of their manuscript when sending the final version to the editorial office.

Conditions

All manuscripts are subject to editorial review. Manuscripts are received with the explicit understanding that they are not under simultaneous consideration by any other publication. Submission of a manuscript for publication implies the transfer of copyright from the author to the publisher upon acceptance. Accepted papers become the permanent property of 'Brain, Behavior and Evolution' and may not be reproduced by any means, in whole or in part, without the written consent of the publisher. It is the author's responsibility to obtain permission to reproduce illustrations, tables, etc. from other publications.

Arrangement

The following major sections should each begin a new page of the manuscript:

Title page: The first page of the manuscript should include the title of the paper and the following additional information for publication: the name of each author; the name and address of each author's relevant institution (usually the institution of affiliation at the time the work was done – if different, an author's current institution and address can be indicated by footnote); and a running title for page headings. The number of figures and tables should also be indicated for the journal office, as should the corresponding author's complete mailing and delivery address(es), e-mail address, telephone number, and Fax number.

Key words: A list of 3–10 key words should be included for indexing purposes. Key words used in the title must be repeated here.

Abstract: The abstract should be restricted to 1 page and should summarize the entire paper, including the results. The term 'will be discussed' should be avoided, as should abbreviations and citations.

Text: The text of the manuscript should comprise the following sections: Introduction, Materials and Methods, Results, and Discussion. Within each section, successive levels of subheadings (if used) should be clearly distinguished. Acknowledgments (if appropriate) should be added at the end of the text. Footnotes should be avoided; if essential, they should be numbered consecutively and typed at the bottom of the appropriate page. Citations in the text should be enclosed in [brackets], not (parentheses), and should comprise the name(s) of the author(s) – use 'et al.' when there are three or more – and the year of publication, in chronological order, as in the following examples: [Reiner et al., 1984a; Brauth, 1988, 1990; Schwerdtfeger and Smeets, 1988]. (Please note: the first Reiner et al., 1984 reference to be cited is automatically 1984a in that and subsequent citations and in the References – see examples.) Work in preparation is formally cited as '[unpublished observations]', and these are assumed to be those of all (and only) the authors. If otherwise, first initials and last names should be used, as for '[personal communication]' from a colleague, as neither appears in the References. Statistics reported in the text must include the value of the test statistic (F, t, r, etc.), the number of subjects (n) or degrees of freedom (df), and the p value (<, >, or the exact value). They should be written in one of the following forms: $F = 19.44$, d.f. = 6,62, $p < 0.001$ or $F_{(6,62)} = 19.44$, $p < 0.001$; $t = 3.73$, $n = 25$, $p < 0.01$ or $t_{(25)} = 3.73$, $p < 0.01$; $r = 0.34$, $n = 8$, $p > 0.05$ or $r = 0.34$, $n = 8$, n.s. Reported means must always be accompanied by the standard deviation or standard error of the mean and should be written in a form such as 70 ± 2.2 mm.

Tables: Each table should bear a heading incorporating the table number (in Arabic numerals) and title; if necessary, a key to symbols in the table should appear beneath the table.

Figure legends: Figure legends should be numbered and listed in order. If a figure has more than one part, the legend should begin with a short description of the figure as a whole before each individual frame is described or referred to.

Figures: Illustrations for publication must be professional quality drawings or photographs; negatives can not be used. Due to technical reasons, figures with a screen background should not be submitted. Photocopies are sufficient for review purposes only if they allow assessment of the figure's quality. The figure number, author and 'top' should be clearly indicated on the back of each figure with a soft pencil.

Color illustrations: Up to 6 color illustrations per page can be integrated within the text at the reduced price of CHF 660.– / USD 545.00 per page. Color illustrations are reproduced at the author's expense.

Cover illustration: Authors may submit a suggested cover illustration after a paper has been accepted. Suggested illustrations may be black and white drawings or photographs, and must be accompanied by a one-sentence caption.

References: The list of references should be arranged alphabetically (except that two-authored papers are listed before multi-authored papers with the same first author) and should include only those publications cited in the text. Work accepted for publication may be listed as 'in press'. 'Unpublished observations' or 'personal communications' cited in the text are not listed in the references. *The names of all authors and/or editors should appear in each reference.* Periodicals should be abbreviated according to Index Medicus; please punctuate and format references exactly according to the following examples.

Atema, J., R.R. Fay, A.N. Popper, and W.N. Tavolga (eds.) (1988) Sensory Biology of Aquatic Animals. Springer, New York.

Bodznick, D., and R.L. Boord (1986) Electroreception in Chondrichthyes: central anatomy and physiology. *In* Electroreception (ed. by T.H. Bullock and W. Heiligenberg), Wiley, New York, pp. 225–256.

Braford, M.R., Jr. (1971) Efferent Projections of the Corpus Cerebelli in the Goldfish. Ph.D. dissertation, Case Western Reserve University, Cleveland, Ohio.

de Beer, G. (1958) Embryos and Ancestors (3rd ed.). Oxford University Press, London.

Eaton, R.C., and R. DiDomenico (1985) Command and the neural causation of behavior: A theoretical analysis of the necessity and sufficiency paradigm. Brain Behav. Evol., *27:* 132–164.

Grillner, S. (1981) Control of locomotion in bipeds, tetrapods and fish. *In* Motor Control (ed. by V. Brooks), Handbook of Physiology, Section 1, Vol. 2, Part 2, American Physiological Society, Bethesda, pp. 1179–1236.

Kanwal, J.S., and T. E. Finger (1989) Convergence of gastric and gustatory inputs in the medulla of the channel catfish, *Ictalurus punctatus.* Soc. Neurosci. Abstr., *15:* 929.

Lohman, A.H.M., P.V. Hoogland, and R.J.G.M. Witjes (1988) Projections from the main and accessory olfactory bulbs to the amygdaloid complex in the lizard *Gekko gecko. In* The Forebrain of Reptiles (ed. by W.K. Schwerdtfeger and W.J.A.J. Smeets), Karger, Basel, pp. 41–49.

Reiner, A., and H.J. Karten (1985) Comparison of olfactory bulb projections in pigeons and turtles. Brain Behav. Evol., *27:* 11–27.

Reiner, A., B.M. Davis, N.C. Brecha, and H.J. Karten (1984a) The distribution of enkephalin-like immunoreactivity in the telencephalon of the adult and developing domestic chicken. J. Comp. Neurol., *228:* 245–262.

Reiner, A., J.E. Krause, K.T. Keyser, W.D. Eldred, and J.F. McKelry (1984b) The distribution of substance P in turtle nervous system: a radioimmunoassay and immunohistochemical study. J. Comp. Neurol., *226:* 50–75.

Schwerdtfeger, W.K., and W.J.A.J. Smeets (eds.) (1988) The Forebrain of Reptiles. Karger, Basel.

Senn, D.G. (1968) Bau und Ontogenese von Zwischen- und Mittelhirn bei *Lacerta sicula* (Rafinesque). Acta Anat., *65:* Suppl. 55.

ten Donkelaar, H.J., and R. de Boer-van Huizen (1981) Basal ganglia projections to the brain stem in the lizard *Varanus exanthematicus* as demonstrated by retrograde transport of horseradish peroxidase. Neuroscience, *6:* 1567–1590.

KARGER

Fax + 41 61 306 12 34
E-Mail karger@karger.ch
www.karger.com

The Guidelines for Authors are available at:
www.karger.com/journals/bbe/bbe_gl.htm

Vertebrate animals
If vertebrate animals are involved, the author should complete and include in the Materials and Methods section a statement similar to the following example. 'The original research reported herein was performed under guidelines established by ... .'

Further suggestions for the preparation of manuscripts: Abbreviations should not be over-used in the text. In any case, a term should be written out the first time it is used and the abbreviation established in parentheses. Abbreviations should never be used to begin a sentence. Single authors should avoid the use of the first person plural, and all authors should give careful thought to the use of tense. For example, information may flow more smoothly to the reader if the present tense is reserved for established phenomena and the past tense used for the results being reported. Similarly, the active voice is usually preferable to the passive voice. Authors should also give thought to exact word usage. Taxonomic names and precise biological terms should be used to refer to animal groups, and vague, subjective descriptors such as 'higher' and 'lower' should be avoided. Similarly, the descriptive prefixes paleo-, archi-, and neo- applied to cerebellum, cortex, or striatum imply an invalid phylogenetic sequence and should also be avoided. In addition, authors should refrain from the use of invalid typological concepts. Instead of 'the fish', a more accurate and specific category is preferable: 'the channel catfish', 'silurid catfishes', 'all teleosts examined to date', or 'most actinopterygian fishes', for example. Authors are also reminded that, by definition, there are no living primitive animals, only living animals with features that are either primitive or derived, which can be determined only by appropriate phylogenetic analysis.

Page charges
There are no charges for papers of 10 or fewer printed pages (including tables, illustrations and references). Each additional complete or partial page is charged to the author at CHF 275.– / USD 196.00 if the author's institution does not carry a current institutional subscription to the journal. As the page charges are waived in any case for the first paper each year to be published by a member of The J.B. Johnston Club, authors who are members of JBJC should so indicate on the title page.

Galley proofs
Unless otherwise requested, galley proofs will be sent to the corresponding author and address indicated on the title page. Authors should plan to correct and return galley proofs with the least possible delay. Excessive alterations made in galley proofs, other than correction of printer errors, may be charged to the author. No page proofs are supplied.

Reprints
Order forms and a price list are sent with the galley proofs. Orders submitted after the issue is printed are subject to considerably higher prices. Upon request, reprints can be sent by airmail at the author's expense. An author who has a personal subscription to the journal may request 50 reprints gratis, regardless of whether additional reprints are ordered. The request should be noted on the reprint order form and should be accompanied by a photocopy of the paid invoice or other proof of a current personal subscription in the author's name.

General Information

Brain, Behavior and Evolution

Publication data: 'Brain, Behavior and Evolution' is published 12 times a year. Volumes 59, 60 with 6 issues each appear in 2002.

Subscription rates: Subscriptions run for a full calendar year. Prices are given per volume.
Personal subscription:
Subscription + postage and handling
CHF 526.50 + CHF 28.80 Europe
CHF 526.50 + CHF 42.70 Overseas
USD 405.00 + USD 28.80
+ online subscription per volume:
CHF 39.–/USD 30.00
(Must be in the name of, billed to, and paid by an individual. Order must be marked 'personal subscription'.)
Institutional subscription:
Subscription + postage and handling
CHF 1053.– + CHF 36.– Europe
CHF 1053.– + CHF 53.40 Overseas
USD 810.00 + USD 36.00
+ online subscription per volume:
CHF 58.–/USD 45.00
Airmail postage: CHF 36.–, USD 30.00.
(Extra per volume.)
Discount subscription prices:
Members of the APA; Int. Brain Research Org.:
Subscription + postage and handling
CHF 526.50 + CHF 28.80 Europe
CHF 526.50 + CHF 42.70 Overseas
USD 405.00 + USD 28.80
J.B. Johnston Club; Int. Soc. for Neuroethology; Soc. for Integrative and Comparative Biology:
Subscription + postage and handling
USD 183.00 + USD 12.00
Students:
Subscription + postage and handling
USD 99.75 + USD 12.00

Microforms: Please direct all enquiries and orders to: University Microfilms Inc.
300 North Zeeb Road, P.O. Box 1346
Ann Arbor, MI 48106-1346 (USA)

Subscription orders: Orders can be placed at agencies, bookstores, directly with the Publisher
S. Karger AG, P.O. Box,
CH–4009 Basel (Switzerland)
or with any of the following distributors:
Deutschland: S. Karger GmbH, Postfach
D–79095 Freiburg
France: Librairie Luginbühl
36, bd de Latour-Maubourg, F–75007 Paris
Great Britain, Ireland: S. Karger AG
4 Rickett Street
London SW6 1RU, UK
USA: S. Karger Publishers, Inc.
26 West Avon Road, P.O. Box 529
Farmington, CT 06085, USA
India, Bangladesh, Sri Lanka:
Panther Publishers Private Ltd.
33, First Main
Koramangala First Block
Bangalore 560 034, India
Japan: Karger Japan, Inc.
Yushima S Bld. 3F
4-2-3, Yushima, Bunkyo-ku
Tokyo 113-0034, Japan

Change of address: Both old and new address should be sent to the subscription source.

Single issues and back volumes: Information can be obtained through the Publisher.

Advertising: Correspondence and rate requests should be addressed to the Publisher.

Bibliographic indices: This journal is regularly listed in bibliographic services, including *Current Contents®*.

KARGER

Fax +41 61 306 12 34
E-Mail karger@karger.ch
www.karger.com

The Journal Home Page is available at:
www.karger.com/journals/bbe

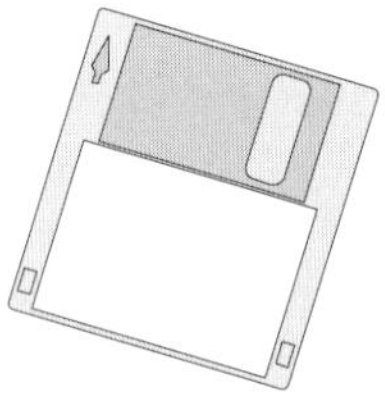

Disk Submission

A **comprehensive version** of these guidelines for the **submission of manuscripts and illustrations on disk** can be accessed online under **www.karger.com/journals/disksub.htm** or requested as a leaflet from the editorial department.

Please follow the general rules on style and arrangement in the **'Guidelines for Authors'** of the journal. Submit a **double-spaced printout** of your manuscript matching the file on disk exactly.

Technical Information
Store only data pertinent to this manuscript. Include your fonts (screen and printer fonts for Macintosh). For a **text only** manuscript, use a 3.5-inch disk. For **text and illustrations,** use 3.5 or 5.25 Magneto Optical Disc; CD-ROM (DOS or hybrid); iomage Zip/DOS (100 MB), or 44 MB, 88 MB or 200 MB SyQuest. Compress extensive data using ZIP format.

Data Description
Label your disk with the **title of the journal or book** for which it is submitted; a **short version** of the paper's title; your **name,** postal and e-mail address, fax and phone number; **hard- and software** used, including the version; and the **complete list of the file names.** File names consist of eight characters before the dot identifying the file and three characters after the dot identifying the application.

Text and Format
The **preferred word-processing package is MS-Word.** Other commonly used PC text programs and ASCII formats are accepted.
Update your data when you revise subsequent to the reviewer's or editor's decision and mark revisions clearly. Manuscripts accepted for publication are subject to **copy editing:** changes and/or questions will be marked in the manuscript, which will be returned to you with your disk and the galley proofs for checking.
Enter your text continuously flush left. Do not use any layout functions. Do not split words at the end of a line. **Do not use indentations.** Use a blank line before and a hard return at the end of paragraphs.
Arrange headings flush left. Insert only **one space after punctuation marks.** Insert Greek letters, mathematical symbols etc. with your word-processing program. Use **boldface and italics** as well as sub- and superscript where appropriate.
Tables are part of the text and should be placed at the end of the text file. Separate the individual parts of a table by lines. Use the tabulator instead of spaces. Format operation data on one line in the same column. **Legends to illustrations are part of the text** and should also be listed at the end of the text file.

Illustrations
Store illustrations as separate files: do not integrate them in your text. Export **line drawings/vector graphics** as TIF, EPS or WMF format. Use Photoshop for processing and retouching **scanned halftone images.** Save the original scan and the processed version. Export **black and white or color images** as TIF or EPS format in their anticipated size in print.
For detailed information on how to submit illustration data on disk please consult the leaflet or **www.karger.com/journals/disksub.htm.**

KI 98 DISK

KARGER

Fax +41 61 306 12 34
E-Mail karger@karger.ch
www.karger.com

The Journal Home Page is available at:
www.karger.com/journals/bbe

Brain, Behavior
and Evolution

Evolutionary Convergence as a Tool in Neuroscience

13th Annual Karger Workshop
San Diego, Calif., November 9, 2001

Guest Editors
Heather L. Eisthen, East Lansing, Mich.
Kiisa C. Nishikawa, Flagstaff, Ariz.

30 figures and 1 table, 2002

KARGER

Basel · Freiburg · Paris · London · New York ·
New Delhi · Bangkok · Singapore · Tokyo · Sydney

S. Karger
Medical and Scientific Publishers
Basel · Freiburg · Paris · London
New York · New Delhi · Bangkok
Singapore · Tokyo · Sydney

Drug Dosage
The authors and the publisher have exerted every effort to ensure that drug selection and dosage set forth in this text are in accord with current recommendations and practice at the time of publication. However, in view of ongoing research, changes in government regulations, and the constant flow of information relating to drug therapy and drug reactions, the reader is urged to check the package insert for each drug for any change in indications and dosage and for added warnings and precautions. This is particularly important when the recommended agent is a new and/or infrequently employed drug.

Printed in Switzerland on acid-free paper by
Reinhardt Druck, Basel
ISBN 3–8055–7481–9

KARGER

Fax + 41 61 306 12 34
E-Mail karger@karger.ch
www.karger.com

Vol. 59, No. 5–6, 2002

Brain, Behavior and Evolution

Contents

KARGER

Fax + 41 61 306 12 34
E-Mail karger@karger.ch
www.karger.com

Access to full text and tables of contents, including tentative ones for forthcoming issues:
www.karger.com/journals/bbe/bbe_bk.htm

Brain, Behavior and Evolution
Brain Behav Evol 2002;59:235–239

Convergence: Obstacle or Opportunity?

Heather L. Eisthen[a] Kiisa C. Nishikawa[b]

[a]Department of Zoology, Michigan State University, East Lansing, Mich., and [b]Department of Biological Sciences, Northern Arizona University, Flagstaff, Ariz., USA

Evolutionary convergence is a remarkable phenomenon that spans all levels of biological organization. Examples are ubiquitous, and often involve correlated changes in molecules, morphology, physiology, and behavior. Birds and mammals convergently evolved the ability to regulate their body temperature, an adaptation that involves a suite of changes in physiology and anatomy, and has enabled both groups to expand into niches not available to many exothermic animals. Bats, birds, and some insects have acquired flapping flight independently, and again some of the underlying changes and resulting abilities are remarkably similar. Some bats, dolphins, and teleost fishes send out signals that produce echoes used to localize objects in the environment. What can we learn about neurobiology and behavior by investigating such phenomena?

Light-detecting eyes have evolved repeatedly in different animal phyla. Many elements of eye structure are convergent; for example, lenses have evolved repeatedly, presumably because the ability to form an image can be adaptive for many organisms inhabiting a variety of niches. Similarly, the ability to discriminate different wavelengths of light reflected by objects in the environment (color vision) has evolved repeatedly, perhaps as an adaptive trait that allows for improved selection of food items or more elaborate intraspecific communication [Pichaud et al., 1999]. The repeated origin of these features provides opportunities for us to test hypotheses concerning the pressures that have given rise to these features, and the function of each. Are these examples of convergence due to functional demands, mechanical constraints related to the physics of light, or the repeated deployment of a network of developmental genes? Are these three explanations of convergence mutually exclusive? The papers contained in this volume explore these questions through detailed analysis of a wide range of examples.

In organizing this workshop and compiling this volume, our goal is to urge comparative neurobiologists to consider the heuristic value of evolutionary convergence (homoplasy) in helping us understand the diversity of structure and function in the nervous system; that is, we want to encourage our colleagues to view convergence as an *opportunity* rather than an *obstacle* [Brooks, 1996]. We contend that a close examination of instances of evolutionary convergence can provide a wealth of information concerning processing strategies used by the nervous system, as well as constraints on neural architecture.

Convergence is a complex and important topic that spans taxa, levels of analysis, and approaches to studying the nervous system. It has been the subject of much recent interest in evolutionary biology, including a volume of papers that address the challenges and opportunities afforded by convergence [Sanderson and Hufford, 1996]. In 'The Crucible of Creation', Simon Conway Morris strives to refute the popular belief that evolutionary processes

KARGER
Fax + 41 61 306 12 34
E-Mail karger@karger.ch
www.karger.com

0006–8977/02/0596–0235$18.50/0

Accessible online at:
www.karger.com/journals/bbe

Heather L. Eisthen
Department of Zoology
Michigan State University, 203 Natural Science Building
East Lansing, MI 48824 (USA)
Tel. +1 517 353 1953, Fax +1 517 432 2789, E-Mail eisthen@msu.edu

produce essentially random patterns, and presses the argument that the appearance of animals with human-like intelligence was inevitable; that, had we not appeared, some other organism would have evolved many of the same behaviors that we display [Conway Morris, 1998]. Although many of the best-known examples of convergence are morphological, convergence occurs at every level of biological organization. In this volume, the authors consider examples of convergence at the level of molecules, physiology, anatomy, and behavior.

The volume begins with an overview of convergence in the nervous system and behavior, provided by a consideration of three striking examples of convergent evolution among closely-related species. Nishikawa [2002] describes evidence that asymmetrical ears have evolved repeatedly among owls, a specialization that is unique among vertebrates. This asymmetry affords greatly improved sound localization, allowing owls to hunt in near-dark conditions. She considers another example well-known to comparative neurobiologists: the repeated origin of electroreception in teleost fishes, and the convergent circuitry and behavior that allow electrolocating fishes to avoid jamming signals from nearby conspecifics. Finally, she describes the repeated evolution of tongue projection in anuran amphibians. Different taxa use different morphological and neural specializations to thrust the tongue forward, but all of these specializations allow frogs to catch prey at a distance. This paper points out the widespread nature of convergence, even among closely-related animals, and raises interesting questions about the mechanisms underlying repeated evolution of similarity within narrow taxonomic groups.

Next, Zakon [2002] tackles the problem of convergence at the molecular level. Following a scheme proposed by Doolittle [1994], he points out that molecules can converge in different ways: they can develop similar functions, mechanisms of action, structures, and/or sequences. Using examples drawn from neurobiology, such as the opsin genes underlying color vision in primates, Zakon considers the roles that different types of convergence may have played in the evolution of molecules. Some of the examples lead to surprising conclusions. When a neurotransmitter has both ionotropic and metabotropic receptors, is the existence of these receptors an example of convergence? New evidence suggests that the ionotropic and metabotropic glutamate receptors share a homologous glutamate-binding region, and that a recombination event involving G protein-coupled receptors led to the subsequent divergence in the structure and function of these two receptor types. Animals like scorpions and cone snails produce large numbers of neurotoxins, leading Zakon to suggest that a convergent *strategy* of high rates of gene duplication and recombination might govern the evolution of large gene families.

The organization and evolution of mammalian cortex is a subject of perennial interest in comparative neurobiology, despite (or perhaps because of) severe limitations imposed by availability of material. Kaas [2002] describes the organization of visual cortex in primates and their near relatives, tree shrews. Tree shrews are often used as an outgroup for comparison with primates, and features found in tree shrews are frequently assumed to represent the plesiomorphic condition. By comparing tree shrews to grey squirrels, which are not closely related but inhabit a remarkably similar niche, Kaas demonstrates that tree shrews possess many specializations that are probably specific adaptations related to their diurnal, arboreal lifestyle. The insights gained from this comparison illustrate that convergence can be useful even to researchers whose goal is to investigate homologous features, in both obvious and not so obvious ways. Certainly, by examining taxa inhabiting similar niches, researchers can distinguish convergent features, and remove these from consideration as homologues. However, if convergence is as widespread in the nervous system as suggested here, researchers investigating animals that are difficult to study may choose to examine convergent features in ecologically similar taxa for some purposes.

Continuing the theme of convergence in sensory systems, the following two papers analyze examples of convergence in the olfactory and auditory systems. First, Eisthen [2002] analyzes features of olfactory systems in different phyla that appear to present clear examples of convergent evolution, and which have been the subject of much discussion among olfactory neurobiologists: odorant binding proteins, odorant receptor molecules, signal transduction strategies, and glomerular neuropils in the central nervous system. By considering the detailed structure and function of these features, she illustrates that convergent features may arise through a combination of adaptation and constraint, and that few features constitute 'pure' adaptations. The lesson here is that neurobiologists should not assume that convergent features represent adaptive mechanisms that have arisen in response to a particular problem, and should keep in mind the other forces that may be at work.

The organization and function of the auditory system is considered by Carr and Soares [2002]. They propose that the circuitry that underlies sound localization in birds and mammals represents convergent evolution

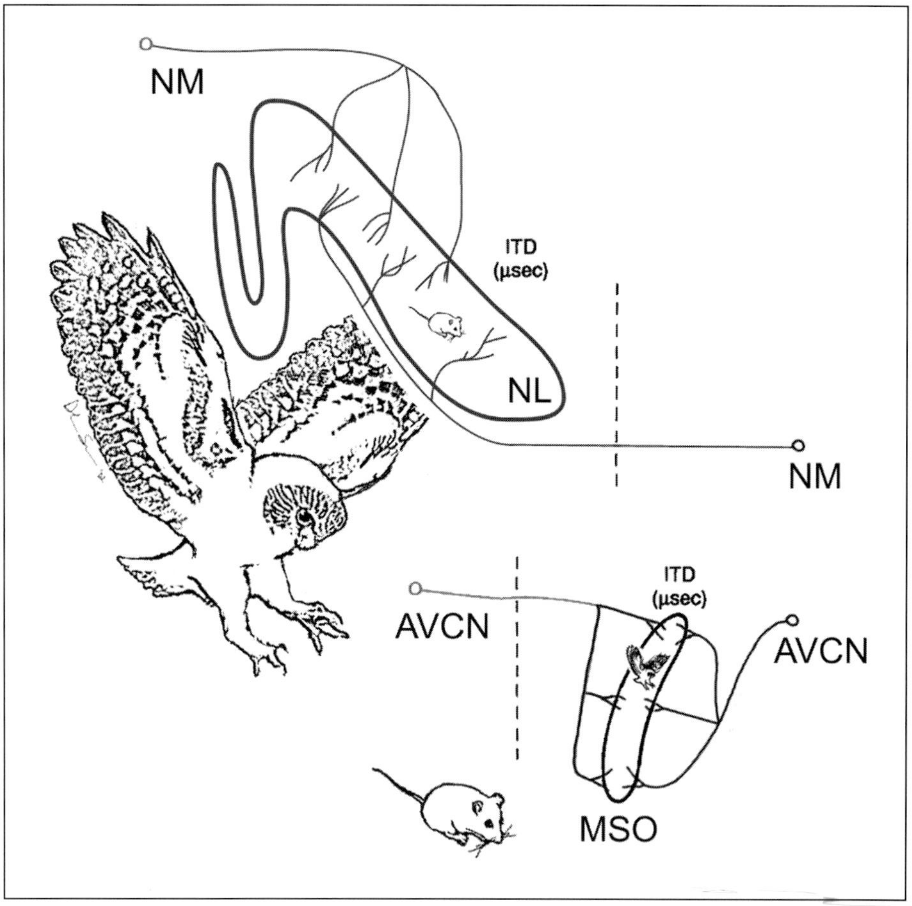

Fig. 1. When hunting, owls use extremely small differences in the time and intensity of sounds arriving at each ear to localize a sound source, such as a mouse. Mice and other mammals use similar circuitry and computations to localize sounds, such as those produced by the flapping wings of an approaching owl. New data suggest that the mechanisms underlying time coding have evolved convergently in birds and mammals [see Carr and Soares, 2002]. AVCN = Anterior ventral cochlear nucleus; ITD = interaural time difference; MSO = medial superior olive; NL = nucleus laminaris; NM = nucleus magnocellularis.

(fig. 1), rather than reflecting underlying homology, as is commonly assumed. Their hypothesis rests on two lines of evidence: the probability that tympanic ears evolved independently in the ancestors of birds and mammals, as described by Clack in a previous Karger Workshop [Clack, 1997]; and new data demonstrating that the cellular interactions involved in precise time coding, which underlies sound localization, are different in birds and mammals. It is astonishing that fine specializations, such as the shedding of dendrites in favor of a calyx-like terminal that envelops the soma of the postsynaptic cell, can evolve repeatedly to perform the same function in different animals, and should cause us to question many of our assumptions about homology of circuitry in different taxa.

Bell [2002] takes the analysis a step further in his consideration of cerebellum-like structures in the hind- and midbrain of all classes of vertebrates. The cerebellum, octavolateral nuclei in some aquatic vertebrates, electrosensory lobes in some electric fish, dorsal cochlear nucleus in therian mammals, and other structures display a similar organization in which sensory information is mapped topographically in a deep layer and Purkinje-like cells that make synapses in this deep layer interact with a more superficial array of fibers that are oriented in parallel. The general function of these structures seems to be to monitor sensory input related to motor activity, and to subtract predictable signals produced by the animal's own motor output. Such structures are present in all vertebrates, and some groups have many such structures, prompting Bell to exclaim in his talk that 'mormyrids are a cerebellum festival.' Interestingly, all these structures arise developmentally from the alar plate and are found in mid- and hindbrain, leading Bell to suggest that similar developmental mechanisms, including similar genetic networks, must underlie the development of cerebellum-like structures. If so, does this mean that these structures are, on some level, homologous?

The answer to this question is provided by Wray [2002]. In his paper, Wray takes a 'ground-up' approach in which he considers examples provided by developmental regulatory genes that are expressed in a similar fashion in diverse animals, and then asks whether the resulting structures are homologous. For example, the gene distal-less is expressed repeatedly in tips of appendages and other structures that protrude from the body; it is clear that the tube feet of echinoderms, the limbs of vertebrates and arthropods, and the siphons of ascidians are not homologous, despite their use of distal-less during development. A simple explanation is that such developmental mechanisms might function like modules that can be deployed at different times in different places to produce particular types of structures, and that we cannot infer homology based simply on similar patterns of gene expression. This conclusion has important implications for comparative neurobiologists, and should be considered when assessing ideas such as the popular claim that eyes in insects and vertebrates are homologous because the *Pax-6* gene plays an important role in their development [Harris, 1997]. In addition, Wray considers the evolution of life-history strategies in sea urchins, and demonstrates that the adoption of a non-feeding larval stage has led repeatedly to similar larval morphology, at least among closely-related animals. Overall, the examples presented in this paper demonstrate that convergence can affect development at all levels of organization, and that similarities in aspects of development should not be given special weight when trying to assess homology.

A few overarching themes emerge from the papers presented in this volume. The most obvious is that convergent features arise in response to similar processing demands, and that many features of the nervous system that are assumed to be homologous at first glance may prove to be convergent on closer examination. Convergent mechanisms may be easier to recognize when the functional demands are well understood: many of the papers in this volume discuss examples of convergence in sensory systems, where the input and function are clear. Similar structures can evolve independently even among very closely-related animals, perhaps in part due to the availability of similar substrates on which similar developmental mechanisms can operate. Finally, convergence operates at all levels of biological organization, and features that are convergent at one level might reflect pressures operating at another level.

Convergence is not simply the inverse of homology, an annoyance that creates confusion when trying to discern patterns of homology in comparative studies. Convergence can also provide enlightenment by drawing our attention to features of the nervous system that are associated repeatedly with specialized functions, suggesting mechanisms by which the nervous system processes information and produces behavior.

We would like to thank the many people who contributed to the success of this symposium, and this volume. Laura Bruce provided a terrific amount of logistical support, before, during, and after the Karger Workshop. Bob and Golda Leonard helped administer and run the meeting this year, and saved the day by appearing on short notice with an LCD projector. Walt Wilczynski provided much advice concerning the symposium and manuscripts, and Blinda McClelland helped enormously with the manuscripts. Daphne Soares produced the lovely figure that accompanies this article. We thank both Harvey Karten and Nick Strausfeld for many lengthy and informative discussions about the importance of convergence in understanding the workings of the nervous system, which helped stimulate this symposium. All the speakers/authors were delightful to work with, and we thank them for their thought-provoking talks and papers and for their prompt delivery of manuscripts after the symposium. We thank Glenn and Mary Sue Northcutt for beginning the tradition of the annual J.B. Johnston Club meetings and Karger Workshops, which always provide a stimulating forum in which comparative neurobiologists can share ideas. Finally, we thank Dr. Thomas Karger for his continuing support of *Brain, Behavior and Evolution*, the J.B. Johnston Club, and the annual Karger Workshops.

References

Bell, C.C. (2002) Evolution of cerebellum-like structures. Brain Behav. Evol., *59:* 312–326.

Brooks, D.R. (1996) Explanations of homoplasy at different levels of biological organization. *In* Homoplasy: The Recurrence of Similarity in Evolution (ed. by M.J. Sanderson and L. Hufford), Academic Press, San Diego, CA, pp. 3–36.

Carr, C.E., and D. Soares (2002) Evolutionary convergence and shared computational principles in the auditory system. Brain Behav. Evol., *59:* 294–311.

Clack, J.A. (1997) The evolution of tetrapod ears and the fossil record. Brain Behav. Evol., *50:* 198–212.

Conway Morris, S. (1998) The Crucible of Creation: The Burgess Shale and the Rise of Animals. Oxford University Press, Oxford, UK.

Doolittle, R.F. (1994) Convergent evolution: The need to be explicit. Trends Biochem. Sci., *19:* 15–18.

Eisthen, H.L. (2002) Why are olfactory systems of different animals so similar? Brain Behav. Evol., *59:* 273–293.

Harris, W.A. (1997) *Pax-6:* Where to be conserved is not conservative. Proc. Natl. Acad. Sci. USA, *94:* 2098–2100.

Kaas, J. (2002) Convergences in the modular and areal organization of the forebrain of mammals: Implications for the reconstruction of forebrain evolution. Brain Behav. Evol., *59:* 262–272.

Nishikawa, K.C. (2002) Evolutionary convergence in nervous systems: Insights from comparative phylogenetic studies. Brain Behav. Evol., *59:* 240–249.

Pichaud, F., A. Briscoe, and C. Desplan (1999) Evolution of color vision. Curr. Opin. Neurobiol., *9:* 622–627.

Sanderson, M.J., and L. Hufford (1996) Homoplasy: The Recurrence of Similarity in Evolution. Academic Press, San Diego, CA.

Wray, G.A. (2002) Do convergent developmental mechanisms underlie convergent phenotypes? Brain Behav. Evol., *59:* 327–336.

Zakon, H. (2002) Convergent evolution at the molecular level. Brain Behav. Evol., *59:* 250–261.

Brain, Behavior and Evolution

Brain Behav Evol 2002;59:240–249

Evolutionary Convergence in Nervous Systems: Insights from Comparative Phylogenetic Studies

Kiisa C. Nishikawa

Department of Biological Sciences, Northern Arizona University, Flagstaff, Ariz., USA

Key Words

Animal behavior · Electric fish · Electric communication · Frogs · Neuroethology · Owls · Phylogeny · Prey capture · Sound localization

Abstract

Over the past 20 years, cladistic analyses have revolutionized our understanding of brain evolution by demonstrating that many structures, some of which had previously been assumed to be homologous, have evolved many times independently. These and other studies demonstrate that evolutionary convergence in brain anatomy and function is widespread. Although there are relatively few neuroethological studies in which brain and behavior have be studied within an evolutionary framework, three relatively well studied cases are reviewed here: electric communication among gymnotiform and mormyriform fishes, prey capture among frogs, and sound localization among owls. These three examples reveal similar patterns of brain evolution. First, it is clear that novel abilities have evolved many times independently in taxa whose common ancestors lack these abilities. Second, it is apparent that small changes in neural pathways can lead to dramatic changes in an organism's abilities. Brain evolution at this small scale is quite common. The behavioral importance of small scale changes on one hand, and the pervasiveness of convergent evolution on the other, have several implications for understanding brain evolution. First, similar abilities may be conferred by convergent rather than homologous circuits, even among closely related species. Furthermore, closely related species may use the same information in different ways, or they may use different means to obtain the same information. One reason that convergence is so common in the biological world may be that the evolutionary appearance of novel functions is associated with constraints, for example in the algorithms used for a given neural computation. Convergence in functional organization may thus reveal basic design features of neural circuits in species that possess unique evolutionary histories but use similar algorithms to solve basic computational problems.

Introduction

Over the last 20 years, cladistic studies have revolutionized our understanding of brain evolution by demonstrating that many structures, previously assumed to be homologous, in fact have evolved many times indepen-

KARGER

Fax +41 61 306 12 34
E-Mail karger@karger.ch
www.karger.com

Accessible online at:
www.karger.com/journals/bbe

Kiisa C. Nishikawa
Department of Biological Sciences
Northern Arizona University
Flagstaff, AZ 86011-5640 (USA)
Tel. +1 928 523 9497, Fax +1 928 523 7500, E-Mail Kiisa.Nishikawa@nau.edu

dently [Northcutt, 1984; Nishikawa, 1997]. The most clear-cut examples of convergent evolution involve the appearance of similar features with similar functions in distantly related taxa whose common ancestor lacked the feature. Examples are many, including the evolution of color vision in bees and primates [Pichaud et al., 1999]; the olfactory glomeruli of neopteran insects and vertebrates, which share an impressively detailed suite of morphological and physiological similarities [Strausfeld and Hildebrand, 1999; Eisthen, 2002]; the evolution of electric organs and electric communication in gymnotiform and mormyriform fishes [Bullock et al., 1975; Kawasaki, 1993]; the evolution of a variety of cerebellum-like structures among vertebrates [Bell, 2002]; expansion of the hippocampus in food-storing corvid [Basil et al., 1996] and parid [Healy and Krebs, 1993] birds; hypoglossal afferents in the tongues of frogs [Nishikawa, 1999], birds [Wild, 1990] and mammals [Lowe, 1981]; the evolution of song learning in oscine birds and parrots [Nottebohm, 1972; Striedter, 1994]; and the evolution of divisions of the mammalian sensory isocortex, such as area MT in primates and carnivores [Northcutt and Kaas, 1995; Kaas, 2002]. More enigmatic are cases in which a feature appears repeatedly and independently within a relatively restricted taxonomic group. An intriguing example is the case of asymmetrical ears which, during 580 million years of vertebrate evolution, appeared only in owls. Among owls, however, asymmetrical ears evolved at least four and perhaps as many as seven times independently [Norberg, 1977; Volman and Konishi, 1990].

There are numerous cases in which convergent evolution is suspected, but for which definitive information on the presence or absence of the character in the common ancestor is lacking. One example is the auditory neurons of birds and mammals that are responsible for time-coding, which also share an impressive suite of specialized membrane and synaptic properties for coincidence detection [Carr et al., 2001; Carr and Soares, 2002].

It is important to note that, although not contradicting an hypothesis of homology, neither similarity of structure, nor even similarity of developmental regulatory genes [Harris, 1997; Wray, 2002], alone constitutes sufficient evidence for rejecting an hypothesis of convergent evolution. The character in question must also be present in the common ancestor. In contrast, either dissimilarity of structure or absence of the character in the common ancestor is sufficient to reject an hypothesis of homology.

As for most characteristics of organisms, from morphology to molecules [see Zakon, 2002], the evolutionary history of brains as well as the behavior they produce appears to be characterized by pervasive convergent evolution. The frequent occurrence of convergence during brain evolution is due to two factors: high rates of evolution in general [Nishikawa, 1997; Katz and Harris-Warrick, 1999]; and the existence of common selective pressures imposed by fundamental biophysical constraints which, over time, have produced similar neural networks for performing similar computations.

In the vertebrate brain, relatively few major new structures have appeared during more than 580 million years of evolution [Nishikawa, 1997]. Likewise, in invertebrates, identified neurons and the circuits they comprise appear to be highly conservative within and even between phyla [Edwards and Palka, 1991; Katz and Harris-Warrick, 1999]. Nevertheless, small changes in neural circuits are common and can lead to profound differences in function [Nishikawa, 1997; Katz and Harris-Warrick, 1999].

At first glance, evolutionary convergence may appear to complicate evolutionary studies of the nervous system by confounding efforts to identify homologous structures. Upon reflection, however, it is clear that convergent evolution can be an important tool for neuroscience, primarily by revealing basic design features of neural systems. Examples include algorithms used by mormyrid and gymnotid fishes for determining the sign of the difference in phase between a fish's own electric organ discharge and a jamming stimulus from another fish [Kawasaki, 1993; Carr and Friedman, 1999] and time coding circuits in birds and mammals [Carr et al., 2001; Carr and Soares, 2002].

The goal of this paper is to review three case studies for which comparative phylogenetic studies have demonstrated convergent evolution of neural substrates and behavior: 1) the evolution of electric organs and electric communication in gymnotiform and mormyriform fishes; 2) the evolution of proprioceptive afferents that modulate prey capture in frogs; and 3) the evolution of asymmetrical ears and sound localization among owls. For each example, the behavior and its neural control are briefly reviewed, a cladogram is given to illustrate the evolution of the behavior within the group, and the evolutionary patterns and general conclusions are discussed.

Electric Organs and Electric Communication in Gymnotiform and Mormyriform Fishes

Weakly electric fishes live in turbid water and possess specialized electric organs that generate electric signals, as well as an electrosensory system that perceives them [Bullock and Heiligenberg, 1986]. Electric signals are used in social communication and to locate stationary and moving objects in the environment. Social signals are used in agonistic and courtship displays, and may provide information about the species, sex, size, maturation, location, distance and individual identity of the sender. Behavioral repertoires include echoes, duets, and other complex interactions [Hopkins, 1988].

Two different types of discharge patterns are produced by electric organs. The wave-type electric organ discharge (EOD) is quasi-sinusoidal, occurs at a constant rate, and is characterized by a frequency spectrum with narrow peaks at the fundamental frequency and the higher harmonics [Bullock, 1970]. Pulse-type discharges are separated by relatively long intervals and have a broader frequency spectrum [Heiligenberg, 1977].

When two wave-type fishes with similar EOD frequencies encounter each other, their electric fields interfere and their ability to electrolocate or communicate deteriorates. To avoid mutual jamming, fish shift their discharge frequencies away from each other to create a larger frequency difference [Bullock et al., 1972]. In *Eigenmannia*, the frequency shift is always in the direction that immediately reduces frequency overlap. Thus, to implement the jamming avoidance response, a fish must be able to detect the direction of the difference in phase between its own signal and that of its neighbor [Rose et al., 1987].

Electric organs evolved independently (fig. 1) in South American knifefishes (Gymnotiformes, Ostariophysi) and African elephantfishes (Mormyriformes, Osteoglossomorpha) [Alves-Gomes et al., 1995; Alves-Gomes, 1999]. Due in large part to their importance in neuroethology, molecular phylogenies are available for both Gymnotiformes [Alves-Gomes et al. 1995; Alves-Gomes, 1999] and Mormyriformes [Sullivan et al., 2000]. Despite its independent origin in mormyriform and gymnotiform fishes, many aspects of electric communication are strikingly similar in the two groups, including the production of wave-type EODs and a jamming avoidance response. A wave-type EOD appears to have evolved in the common ancestor of mormyriforms [Sullivan et al., 2000] and is present in the basal taxon Gymnarchus [Kawasaki, 1993]. Within gymnotiforms, a wave-type EOD appears to have evolved independently in *Sternopygus* and in the common ancestor of *Eigenmannia* and *Apteronotus* (fig. 1) [Alves-Gomes et al., 1995].

Like the wave-type EOD, a jamming avoidance response also evolved independently in *Gymnarchus* and *Eigenmannia* and, despite differences in neural substrates, they use the same algorithm to calculate the direction of the difference in phase between the fish's own signal and a neighboring signal [Kawasaki, 1993]. Both taxa use electrosensory organs to detect the mixture of the fish's own reafferent signal and the exafferent signal, the time course of amplitude and phase modulation is analyzed at each area of the body surface, detection of phase modulation occurs by comparing the phase of the two signals at different areas on the body, and ambiguity is reduced by integrating across all body areas. In *Eigenmannia*, phase comparisons take place in the midbrain torus semicircularis, whereas they take place in the electrosensory lateral line lobe of the medulla in *Gymnarchus* [Kawasaki, 1993].

Even within gymnotiforms, there are interesting similarities and differences among closely related species in the neural circuits that control jamming avoidance. First, although *Sternopygus* lacks a jamming avoidance response (fig. 1), it nevertheless possesses midbrain neurons that can discriminate the direction of the difference in phase between the fish's own signal and an interfering signal, suggesting that the circuitry used for jamming avoidance in *Eigenmannia* and *Apteronotus* may have been coopted from more general-purpose circuitry that *Sternopygus* uses for detection of moving objects and conspecifics in its environment [Rose et al., 1987]. Second, although both *Eigenmannia* and *Apteronotus* (fig. 1) exhibit a jamming avoidance response and similar premotor circuits [Heiligenberg et al., 1996], there are interesting differences in control of the jamming avoidance response between species. In both *Eigenmannia* and *Apteronotus*, a jamming stimulus that is below the fish's own frequency causes individuals of both species to increase their EOD frequency. In contrast, if the jamming stimulus is above the fish's own frequency, *Eigenmannia* will decrease its frequency, but *Apteronotus* will increase it even more. To increase EOD frequency, *Eigenmannia* recruits the diencephalic prepacemaker nucleus, whereas to decrease EOD frequency, the sublemniscal prepacemaker nucleus is inhibited [Metzner, 1993]. *Apteronotus* possesses both pathways, but uses only the diencephalic prepacemaker nucleus to modulate its EOD frequency. In *Apteronotus*, lesions of the sublemniscal prepacemaker nucleus have no effect on the jamming avoidance response [Heiligenberg et al., 1996].

Both wave-type EODs and jamming avoidance responses have evolved convergently in mormyriform and gymnotiform fishes. The similarity in the algorithms used to compute the direction of phase difference between a fish's own and an interfering signal, despite differences in the neural substrates where the calculations are per-

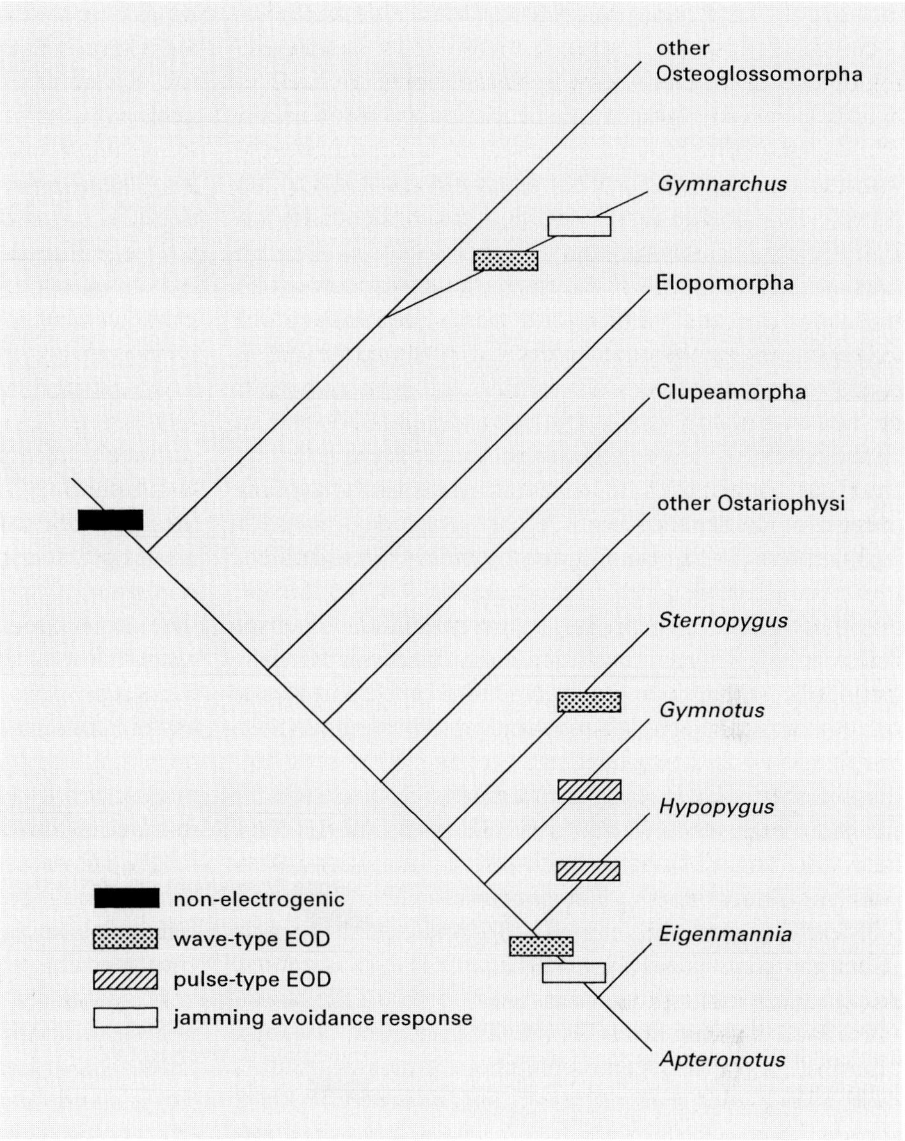

Fig. 1. Cladogram illustrating convergent evolution of wave-type electric organ discharge (EOD) and jamming avoidance responses in African mormyrid and South American gymnotid fishes [based on Alves-Gomes et al., 1995]. The common ancestor of mormyrids and gymnotids was non-electrogenic. Wave-type EODs and jamming avoidance responses evolved independently in *Gymnarchus* and in the common ancestor of *Eigenmannia* and *Apteronotus*. *Sternopygus* possesses a wave-type EOD but lacks a jamming avoidance response.

formed, suggests that common selective pressures imposed by fundamental biophysical constraints have produced similar neural networks in the hindbrain of mormyrids and the midbrain of gymnotids for performing this computation. The diversity in behavior and modulation of EOD frequency within closely related gymnotiform species also suggests that relatively small changes in existing pathways can lead to larger changes in behavior, and that similar circuits may be present, but are nevertheless used for different purposes, in closely related species [Nishikawa, 1997; Katz and Harris-Warrick, 1999; Metzner, 1999].

Prey Capture in Frogs

All frogs possess relatively simple tongues that contain only two pairs of muscles, which protract and retract the tongue during prey capture [Mallett et al., 2001]. Three different mechanisms for protracting the tongue (mechanical pulling, inertial elongation, and hydrostatic elongation) have evolved among frogs. Each is characterized by different biomechanical properties and different mechanisms of neuromuscular control [Nishikawa, 1999]. Frogs primitively possess short tongues that are protracted by mechanical pulling (fig. 2). In mechanical pullers, the tongue shortens to ~ 60–70% of resting length during pro-

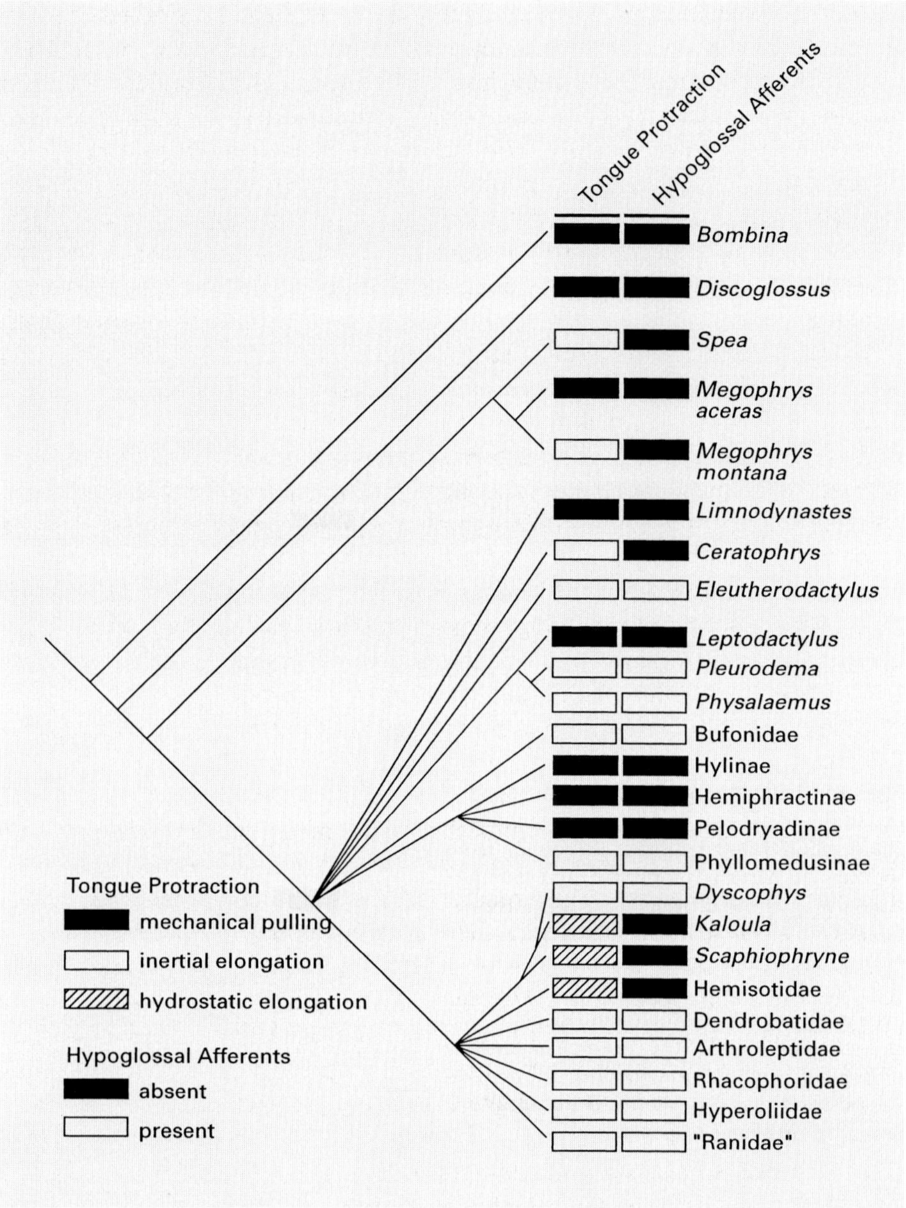

Fig. 2. Cladogram illustrating convergent evolution of ballistic tongue projection (via inertial elongation) and hypoglossal afferents among anurans [based on Ford and Cannatella, 1993; Nishikawa, 1999]. The common ancestor of anurans used mechanical pulling to protract its tongue and lacked hypoglossal afferents. Inertial elongation and hypoglossal afferents that modulate jaw muscle coactivation evolved up to five times independently among anurans.

traction as the protractor muscles contract, pulling the tongue pad forward [Deban and Nishikawa, 1992]. Movements of the head, jaws and tongue are relatively small, slow and asynchronous but tongue movements can be corrected in progress within a single gape cycle. Forward dynamic models of tongue projection in frogs demonstrate that the movement is dynamically stable, and that more than 90% of the force for tongue movement comes from transfer of angular momentum from the opening jaws to the tongue [Mallett et al., 2001].

Ballistic tongue projection has evolved as many as eight times independently in anurans from ancestors that used mechanical pulling to protract the tongue (fig. 2). During ballistic tongue projection, momentum transferred to the tongue from the opening jaws produces tongue elongation of up to 210% of resting length. The tongues of inertial elongators are morphologically similar to those of mechanical pullers, except that relative tongue mass is smaller and the fibers of the tongue protractor and retractor muscles are relatively longer in inertial elongators.

Inertial elongation involves extremely rapid acceleration of the jaws and tongue at rates approaching 600 m/s^2. Our current research focuses on how frog jaw muscles achieve the high in vivo strain rates that are necessary to propel the tongue out of the mouth at high speeds. It appears that the depressor mandibulae muscles and/or their connective tissues store elastic strain energy prior to movement and that recovery of some of this energy amplifies the power of mouth opening [Lappin et al., 2002].

Afferent neurons with axons that travel to the brain in the peripheral hypoglossal nerve have evolved convergently as many as 4–5 times among anurans with ballistic tongue projection (fig. 2). These sensory neurons innervate epithelial mechanoreceptors of the tongue pad [Harwood and Anderson, 2000]. We hypothesize that these mechanoreceptors play a role in coordinating activity of antagonistic muscles during energy storage prior to mouth opening, and in triggering recovery of stored energy during mouth opening. In all anurans that possess them, hypoglossal afferents modulate the phase of activity of the mouth opening and closing muscles [Nishikawa and Gans, 1992; Anderson and Nishikawa, 1993; Monroy et al., 2001]. In intact frogs, the mouth opening muscles are activated before the mouth-closing muscles, whereas in deafferented frogs, mouth opening and closing muscles are activated simultaneously. Thus, when deafferented frogs attempt to feed, their mouths remain closed. By modulating coactivation of the antagonistic mouth opening and closing muscles, hypoglossal afferents likely control the storage and recovery of elastic strain energy by the mouth opening muscles, which boosts the power of mouth opening and tongue elongation [Lappin et al., 2002].

Vertebrates primitively possess purely motor hypoglossal nerves [Ariens-Kappers et al., 1936], although hypoglossal afferents have evolved independently in a number of vertebrate lineages including birds [Wild, 1990] and mammals [Lowe, 1981] as well as among frogs. It is interesting that the embryonic source of hypoglossal afferents differs among anuran lineages [Nishikawa, 1999]. In toads (family Bufonidae), hypoglossal afferents are re-routed glossopharyngeal afferents that ascend to higher brain centers in the solitary tract. In frogs (family Ranidae), they are re-routed cervical spinal afferents that ascend in the dorsomedial funiculus and project to both the granular layer of the cerebellum and the medial reticular formation [Anderson and Nishikawa, 1997].

Ballistic tongue projection has evolved numerous times independently among frogs, as have mechanosensory afferents that modulate the activity of antagonistic jaw muscles. It seems likely that afferent modulation of muscle coactivation boosts the power of mouth opening and tongue elongation by optimizing muscle coactivation for storage and release of elastic strain energy. Toads use solitary afferent pathways for this purpose, whereas frogs use spinal afferents. The function of these afferents appears to be quite similar in the different taxa, which may imply a functional constraint. Anatomically, the circuits are quite different and we are currently exploring physiological differences that might affect function. As in the case of electric fishes, relatively small changes in existing afferent pathways have led to the emergence of a novel function, modulation of the activity of antagonistic jaw muscles in inertial elongators. Likewise, convergent evolution of hypoglossal afferents among anurans with ballistic tongue projection is based on small changes in the peripheral and central connections of sensory neurons that modulate jaw muscle activity.

Sound Localization in Owls

Among vertebrates, only owls have evolved asymmetrical ears. Asymmetrical ears function to determine the vertical position of a sound source [Moiseff, 1989]. Owls with symmetrical ears are reluctant even to fly in total darkness, whereas owls with asymmetrical ears readily capture prey in total darkness [Payne, 1971]. Whereas cladograms are available for genera of weakly electric fishes [Alves-Gomes, 1999; Sullivan et al., 2000] and for families of anurans [Ford and Cannatella, 1993], less information is available concerning the phylogenetic relationships of owl genera [Randi et al., 1991; Rebholz et al., 1993]. To analyze the evolution of sound localization in owls, ear asymmetry data from Norberg [1977] were mapped onto a phylogeny (fig. 3) based on available allozyme and karyological data [Randi et al., 1991; Rebholz et al., 1993].

According to Norberg [1977], *Athene*, some *Bubo*, some *Strix*, all *Otus*, and all *Ninox* have symmetrical ears, whereas *Aegolius*, some *Bubo*, some *Strix*, all *Asio*, and all *Tyto* have asymmetrical ears. If one looks only at the evolution of asymmetrical ears, the hypothesis that ear asymmetry was lost repeatedly among owls is equally parsimonious with the hypothesis that asymmetrical ears were acquired repeatedly among owls. Thus, it is not possible to say definitively whether asymmetrical ears evolved once in the common ancestor of owls and were lost several times independently, or alternatively, whether ear asym-

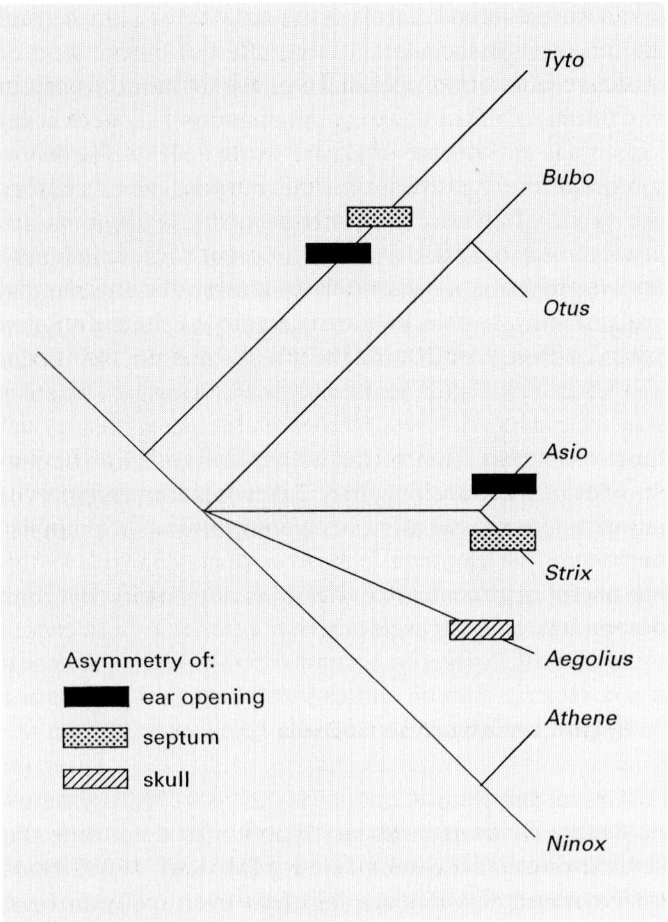

Fig. 3. Cladogram illustrating convergent evolution of ear asymmetry among owls [based on Norberg, 1977; Randi et al., 1991; Rebholz et al., 1993]. *Tyto* and *Asio* possess asymmetries of the soft tissues of the ear opening, and *Tyto* and *Strix* possess asymmetries of the septum. Only *Aegolius* possesses asymmetries of the cranium. Intrageneric variation (not shown) is present in *Bubo* and *Strix*. Given this level of variation in ear anatomy, ear asymmetry may have evolved as many as seven times independently among owls.

metry evolved convergently among owls. However, as noted by Norberg [1977], ear asymmetry differs anatomically among owl lineages (fig. 3). *Tyto*, some *Bubo*, and some *Asio* possess asymmetries in the soft tissues of the ear opening, *Tyto* and some *Strix* possess an asymmetrical septum within the ear opening, and *Aegolius* are unique in possessing an asymmetrical skull. Therefore, the most parsimonious hypothesis is that asymmetrical ears evolved at least four times and perhaps as many as seven times independently among owls. This fact is amazing, considering that no other vertebrates have evolved asymmetrical ears.

All terrestrial vertebrates use binaural cues to localize sounds in space [Fay, 1988]. Species with symmetrical ears, including humans, use both interaural time difference (ITD) and interaural intensity difference (IID) to encode the azimuth of a sound source. In contrast, owls with asymmetrical ears use ITD to encode sound azimuth and IID to encode sound elevation [Volman, 1994]. At the behavioral level, differences in ear morphology and sensory processing in owls with asymmetrical ears translate into the ability to capture prey in total darkness using acoustic cues alone, whereas owls with symmetrical ears are reluctant to fly in total darkness. Neuroanatomically, however, relatively few differences in auditory pathways are responsible for the behavioral difference.

In birds, differences in arrival times of a sound at the two ears are computed by coincidence-detecting neurons in the brainstem nucleus laminaris, which receive delay-line inputs from phase-locked cochlear neurons [Carr and Konishi, 1990]. The dendrites and synaptic channels of laminaris neurons are highly specialized for detecting sub-microsecond time differences [Carr et al., 2001]. Laminaris neurons also translate the time code (phase-locked spikes from the cochlear nucleus magnocellularis) into a place code, in which ITD is represented as the position of neurons within the nucleus magnocellularis [Carr and Friedman, 1999]. This same basic organization of the nucleus laminaris is found in chickens as well as owls, and presumably represents the plesiomorphic condition. The major difference in the organization of the nucleus laminaris between chickens and owls is that in the chicken, the place-map is a monolayer of approximately 1,000 laminaris neurons with a single representation of ITD, whereas barn owls *(Tyto)* possess approximately 10,000 laminaris neurons and multiple representations of ITD [Carr and Friedman, 1999].

Likewise, there are relatively few differences in the organization of the inferior colliculus in owls with symmetrical vs. asymmetrical ears [Volman and Konishi, 1990]. In the midbrain inferior colliculus, maps of auditory and visual space are in register. In both great horned owls (symmetrical ears) and barn owls (asymmetrical ears), the auditory and visual maps of azimuth are oriented anteroposteriorly in the inferior colliculus. In both species, inferior colliculus neurons are tuned to interaural time (ITD) and intensity differences (IID) independent of frequency. In fact, the broad frequency tuning of inferior colliculus neurons appears to represent convergence of frequency-tuned neurons, which may eliminate ambiguities in the interpretation of ITD and IID. In both species, neurons in the inferior colliculus are tuned to either ITD or IID, but not both. The chief differences between great horned owls and barn owls are in the elevation selectivity and narrower tuning of inferior colliculus neurons to interaural intensity differences [Volman and Konishi, 1990].

In summary, asymmetrical ears have evolved numerous times independently among owls. Similar auditory pathways are present not only among owls but also among birds generally. However, the auditory pathways are used for different purposes in different species. For example, all owls possess neurons that are selective for interaural intensity differences (IID) and all owls use this information for sound localization. However, owls with symmetrical ears use IID as an additional azimuth cue, whereas owls with asymmetrical ears use it as an elevation cue. In general, brain nuclei involved in the processing of binaural auditory cues are also similar between owls with symmetrical and asymmetrical ears, except that owls with asymmetrical ears have both greater acuity of sound localization and more neurons devoted to auditory processing. In this system, as in the others, small changes in existing neural pathways appear to underlie the emergence of novel abilities, such as the ability to catch prey in total darkness using auditory cues alone.

Conclusions

Collection of neuroanatomical, neurophysiological, and neuroethological data can be extremely time-consuming even for a single species. Thus, it is not surprising that neuroethology has relatively few comparative case studies in which the anatomical and physiological bases for interesting behavior patterns can be studied within an evolutionary framework [Nishikawa, 1997; Phelps, 2002]. Three relatively well studied examples are electric communication among gymnotiform and mormyriform fishes, prey capture in frogs and ear asymmetry among owls. Given the relative paucity of case studies, it is remarkable that these three examples reveal such similar (not to say convergent) patterns of brain evolution.

First, it is clear that novel behaviors have evolved many times independently in taxa whose common ancestors lack the behavior, from jamming avoidance, to ballistic tongue projection, to nocturnal prey capture. The list of behaviors and brain structures that have evolved convergently is growing rapidly (see Introduction). It is now clear that convergence is as pervasive in nervous systems as it is in the rest of the biological world [Wray, 2002], notwithstanding some claims to the contrary that nervous systems might evolve more conservatively than other biological substrates.

Second, it is apparent that novel behaviors, absent in ancestors, can be conferred by surprisingly small changes in existing neural pathways [Nishikawa, 1997; Katz and Harris-Warrick, 1999]. Why can *Eigenmannia* increase or decrease its EOD to avoid jamming, whereas *Apteronotus* can only decrease it? Why can frogs with inertial elongation open their mouths 18 times faster than mechanical pullers? Why can barn owls capture prey in total darkness whereas great horned owls cannot? Why do humans possess vocabularies on the order of 100,000 words, whereas the most highly trained chimpanzees have difficulty acquiring a vocabulary in excess of 500 'words' [Kandel et al., 2000]? It was long believed that Wernicke's (speech) area in the neocortex was the only new structure to appear in the brain during human evolution [Geschwind, 1965]. However, it now appears that even this structure has a homologue in macaques and other primates [Deacon, 1988]. Small changes in neural pathways can lead to dramatic changes in an organism's abilities, and it is clear that brain evolution at this small scale is quite common.

The behavioral importance of small scale changes on one hand, and the pervasiveness of convergent evolution on the other, have several implications for understanding brain evolution. First, similar abilities may be conferred by convergent rather than homologous circuits, even among closely related species. One example is the proprioceptive afferents that trigger ballistic tongue projection in frogs *(Rana)* and toads *(Bufo)*. Furthermore, closely related species may use the same information in different ways (as in the case of IID-tuned neurons in the inferior colliculus of barn vs. great horned owls), or they may use different ways to obtain the same information. It is difficult to avoid the conclusion that brains are probably more finely tuned to the ecology and natural history of a species than we think.

One reason that convergence is so common in the biological world may be that the evolutionary appearance of novel functions is associated with constraints, for example in the algorithms used for a given neural computation. The time-coding circuits of weakly electric fishes, birds, and mammals are a case in point [Carr and Friedman, 1999]. In all of these sensory systems, timing information is coded by phase-locked spikes and is processed in parallel with other stimulus features. These time-coding circuits exhibit a number of specializations that permit the neurons to resolve sub-microsecond time differences [Kawasaki, 1997; Carr and Friedman, 1999; Carr et al., 2001, Carr and Soares, 2002].

From these case studies, it is clear that comparative phylogenetic analyses have a lot to tell us about the nature of brains and the rules for their transformation.

Acknowledgments

I wish to thank Jenna Monroy, Jason Pilarski and Heather Eisthen for constructive comments on the manuscript. This research was supported by grants from the National Science Foundation (IBN-9809942) and the National Institutes of Health (R25-GM56931).

References

Alves-Gomes, J.A. (1999) Systematic biology of gymnotiform and mormyriform electric fishes: phylogenetic relationships, molecular clocks and rates of evolution in the mitochondrial rRNA genes. J. Exp. Biol., *202:* 1167–1183.

Alves-Gomes, J.A., G. Ortí, M. Haygood, W. Heiligenberg, and A. Meyer (1995) Phylogenetic analysis of the South American electric fishes (Order Gymnotiformes) and the evolution of their electrogenic system: A synthesis based on morphology, electrophysiology, and mitochondrial sequence data. Mol. Biol. Evol., *12:* 298–318.

Anderson, C.W., and K.C. Nishikawa (1993) A prey-type dependent hypoglossal feedback system in the frog, *Rana pipiens*. Brain Behav. Evol., *42:* 189–196.

Anderson, C.W., and K.C. Nishikawa (1997) The functional anatomy and evolution of hypoglossal afferents in the leopard frog, *Rana pipiens*. Brain Res., *771:* 285–291.

Ariens-Kappers, C.U., G.C. Huber, and E.C. Crosby (1936) The Comparative Anatomy of the Nervous System of Vertebrates, including Man. MacMillan, New York.

Basil, J A., A.C. Kamil, R.P. Balda, and K.V. Fite (1996) Differences in hippocampal volume among food storing corvids. Brain Behav. Evol., *47:* 156–164.

Bell, C.C. (2002) Evolution of cerebellum-like structures. Brain Behav. Evol., *59:* 312–326.

Bullock, T.H. (1970) Reliability of neurons. J. Gen. Physiol., *55:* 565–584.

Bullock, T.H., and W. Heiligenberg (eds.) (1986) Electroreception. Wiley, New York.

Bullock, T.H., K. Behrend, and W. Heiligenberg (1975) Comparison of the jamming avoidance response in gymnotoid and gymnarchid electric fish: A case of convergent evolution of behavior and its sensory basis. J. Comp. Physiol. A, *103:* 97–121.

Bullock, T.H., R.H. Hamstra, and H. Scheich (1972) The jamming avoidance response of high frequency electric fish. I. General features. J. Comp. Physiol., *77:* 1–22.

Carr, C.E., and M.A. Friedman (1999) Evolution of time coding systems. Neural Comput., *11:* 1–20.

Carr, C.E., and M. Konishi (1990) A circuit for detection of interaural time differences in the brainstem of the barn owl. J. Neurosci., *10:* 3227–3246.

Carr, C.E., and D. Soares (2002) Evolutionary convergence and shared computational principles in the auditory system. Brain Behav. Evol., *59:* 294–311.

Carr, C.E., D. Soares, S. Parameshwaran, and T. Perney (2001) Evolution and development of time coding systems. Curr. Opin. Neurobiol., *11:* 727–733.

Deacon, T.W. (1988) Human brain evolution. I. Evolution of language circuits. *In* Intelligence and Evolutionary Biology (ed. by H.J. Jerison and I. Jerison), Springer-Verlag, Berlin, pp. 363–381.

Deban, S.M., and K.C. Nishikawa (1992) The kinematics of prey capture and the mechanism of tongue protraction in the green tree frog, *Hyla cinerea*. J. Exp. Biol., *170:* 235–256.

Edwards, J.S., and J. Palka (1991) Insect neural evolution – a fugue or an opera? Sem. Neurosci., *3:* 391–398.

Eisthen, H.L. (2002) Why are olfactory systems of different animals so similar? Brain Behav. Evol., *59:* 273–293.

Fay, R.R. (1988) Hearing in Vertebrates: A Psychophysics Databook. Hill-Fay Associates, Winnetka, IL.

Ford, L., and D.C. Cannatella (1993) The major clades of frogs. Herpetologica, *7:* 94–117.

Geschwind, N. (1965) Disconnection syndromes in animal and man. Part 1. Brain, *88:* 237–294.

Harris, W.A. (1997) *Pax-6*: Where to be conserved is not conservative. Proc. Natl. Acad. Sci., *94:* 2098–2100.

Harwood, D.V., and C.W. Anderson (2000) Evidence for the anatomical origins of hypoglossal afferents in the tongue of the Leopard frog, *Rana pipiens*. Brain Res., *862:* 288–291.

Healy, S.D., and J.R. Krebs (1993) Food storing and the hippocampus in corvids: Amount and size are correlated. Proc. R. Soc. Lond., *248:* 241–245.

Heiligenberg, W. (1977) Principles of Electrolocation and Jamming Avoidance in Electric Fish: A Neuroethological Approach. Springer, New York.

Heiligenberg, W., W. Metzner, C.J.H. Wong, and C.H. Keller (1996) Motor control of the jamming avoidance response of *Apteronotus leptorhynchus*: Evolutionary changes of a behavior and its neuronal substrates. J. Comp. Physiol. A, *197:* 653–674.

Hopkins, C.D. (1988) Neuroethology of electric communication. Ann. Rev. Neurosci., *11:* 497–535.

Kaas, J. (2002) Convergences in the modular and areal organization of the forebrain of mammals: Implications for the reconstruction of forebrain evolution. Brain Behav. Evol., *59:* 262–272.

Kandel, E.R., J.H. Schwartz, and T.M. Jessell (2000) Principles of Neural Science. McGraw-Hill, New York.

Katz, P.S., and R.M. Harris-Warrick (1999) The evolution of neuronal circuits underlying species-specific behavior. Curr. Opin. Neurobiol., *9:* 628–633.

Kawasaki, M. (1993) Independently evolved jamming avoidance responses employ identical computational algorithms: A behavioral study of the African electric fish, *Gymnarchus niloticus*. J. Comp. Physiol. A, *173:* 9–22.

Kawasaki, M. (1997) Sensory hyperacuity in the jamming avoidance response of weakly electric fish. Curr. Opin. Neurobiol., *7:* 473–479.

Lappin, A.K., K.C. Nishikawa, and D.J. Pierotti (2002) A proposed mechanism for high power output during feeding in toads. Am. Zool., *42:* in press.

Lowe, A.A. (1981) The neural regulation of tongue movements. Prog. Neurobiol., *15:* 295–344.

Mallett, E.S., G. Yamaguchi, J.M. Birch, and K.C. Nishikawa (2001) Feeding motor patterns in anurans: Insights from biomechanical modeling. Am. Zool., *41:* 1364–1374.

Metzner, W. (1993) Neural circuitry for communication and jamming avoidance in gymnotiform electric fish. J. Exp. Biol., *202:* 1365–375.

Metzner, W. (1999) Neural circuitry for communication and jamming avoidance in gymnotiform electric fish. J. Exp. Biol., *202:* 1365–1375.

Moiseff, A. (1989) Bi-coordinate sound localization by the barn owl. J. Comp. Physiol. A, *164:* 637–644.

Monroy, J.A., C.W. Anderson, and K.C. Nishikawa (2001) Neuroanatomy of a muscular hydrostatic tongue in microhylid frogs. Soc. Neurosci. Abstr., *27:* 440.

Nishikawa, K.C. (1997) Emergence of novel functions during brain evolution. Bioscience, *47:* 341–354.

Nishikawa, K.C. (1999) Neuromuscular control of prey capture in frogs. Phil. Trans. R. Soc. Lond. B, *354:* 941–954.

Nishikawa, K.C., and C. Gans (1992) The role of hypoglossal sensory feedback during feeding in the marine toad, *Bufo marinus*. J. Exp. Zool., *264:* 245–252.

Norberg, R.A. (1977) Occurrence and independent evolution of bilateral ear asymmetry in owls and implications on owl taxonomy. Phil. Trans. R. Soc. Lond. B, *280:* 375–408.

Northcutt, R.G. (1984) Evolution of the vertebrate nervous system: patterns and processes. Amer. Zool., *24:* 701–716.

Northcutt, R.G., and J.H. Kaas (1995) The emergence and evolution of the mammalian neocortex. Trends Neurosci., *18:* 373–379.
Nottebohm, F. (1972) The origin of vocal learning. Am. Nat., *106:* 116–140.
Payne, R.S. (1971) Acoustic location of prey by barn owls *(Tyto alba)*. J. Exp. Biol., *54:* 535–573.
Phelps, S.M. (2002) Like minds: evolutionary convergence in nervous systems. Trends Ecol. Evol., *17:* 158–159.
Pichaud, F., A. Briscoe, and C. Desplan (1999) Evolution of color vision. Curr. Opin. Neurobiol., *9:* 622–627.
Randi, E., G. Fusco, R. Lorenzini, and F. Spina (1991) Allozyme divergence and phylogenetic relationships within the Strigiformes. Condor, *93:* 295–301.
Rebholz, W.E.R., L.E.M. De Boer, M. Sasaki, R.H.R. Belterman, and C. Nishida-Umehara (1993) The chromosomal phylogeny of owls (Strigiformes) and new karyotypes of seven species. Cytologia, *58:* 403–416.
Rose, G., C. Keller, and W. Heiligenberg (1987) 'Ancestral' neural mechanisms of electrolocation suggest a substrate for the evolution of the jamming avoidance response. J. Comp. Physiol. A, *160:* 491–500.
Strausfeld, N.J., and J.G. Hildebrand (1999) Olfactory systems: Common design, uncommon origins? Curr. Opin. Neurobiol., *9:* 634–639.
Striedter, G.F. (1994) The vocal control pathways in budgerigars differ from those in songbirds. J. Comp. Neurol., *343:* 35–56.
Sullivan, J.P., S. Lavoué, and C.D. Hopkins (2000) Molecular systematics of the African electric fishes (Mormyroidea: Teleostei) and a model for the evolution of their electric organs. J. Exp. Biol., *203:* 665–683.
Volman, S.F. (1994) Directional hearing in owls: Neurobiology, behavior and evolution. *In* Perception and Motor Control in Birds (ed. by M.N.O. Davies and P.R. Green), Springer-Verlag, Berlin, pp. 292–314.
Volman, S.F., and M. Konishi (1990) Comparative physiology of sound localization in four species of owls. Brain Behav. Evol., *36:* 196–215.
Wild, J.M. (1990) Peripheral and central terminations of hypoglossal afferents innervating lingual tactile mechanoreceptor complexes in Fringillidae. J. Comp. Neurol., *298:* 157–171.
Wray, G.A. (2002) Do convergent developmental mechanisms underlie convergent phenotypes? Brain Behav. Evol., *59:* 327–336.
Zakon, H. (2002) Convergent evolution on the molecular level. Brain Behav. Evol., *59:* 250–261.

Brain, Behavior and Evolution

Brain Behav Evol 2002;59:250–261

Convergent Evolution on the Molecular Level

Harold H. Zakon

Section of Neurobiology, University of Texas, Austin, Tex., USA

Key Words

Convergence · Molecular evolution · Na^+ channels · K^+ channels · Opsins · Gap junctions · Glutamate receptors · Pufferfish (Fugu)

Abstract

Divergence and convergence are two evolutionary processes by which organisms become adapted to their environments. With the advent of molecular biological techniques it is possible to ask if these processes are observed at the molecular level. There are many examples of molecular divergence in which molecular sequence or function change over evolutionary time. There are fewer reports of convergent evolution on the molecular level, and these claims are sometimes controversial. In this paper I discuss the types of convergent molecular evolution, describe the criteria for accepting or rejecting convergence, and give some examples relevant to neurobiology where convergence has been claimed. These include convergent evolution of opsins, gap junction proteins, neurotransmitter receptors, ion channels, and venoms directed against ion channels.

Introduction: Convergence and Parallelism

The concept of convergent evolution permeates the study of evolutionary processes in biology. Convergent evolution occurs when a similar-appearing trait evolves independently in two or more lineages from *different* ancestral states. Convergence is often distinguished from parallelism in which a similar trait evolves independently in two lineages from the *same* ancestral state. So, for example, internal fertilization with intromittent organs has evolved convergently in mammals (penis), elasmobranchs (claspers), and some teleosts (gonopodium). These animals do not share a common ancestor with intromittent organs and their intromittent organs develop from different tissues. On the other hand, viviparity, or the internal maturation and nourishment of developing embryos within the oviduct, has appeared multiple times within sharks. In this case, the trait comes from a similar starting point: the elasmobranch oviduct. Viviparity in elasmobranchs is an example of parallel evolution [Pough et al., 1996].

The presence of convergent evolution at the organismal level naturally leads us to ask if convergence occurs on the molecular level. In this paper, I will discuss the forms of and criteria for molecular convergence, and then give examples that pertain to the nervous system.

KARGER

Fax + 41 61 306 12 34
E-Mail karger@karger.ch
www.karger.com

Accessible online at:
www.karger.com/journals/bbe

Harold H. Zakon
Patterson Laboratory, University of Texas
Austin, TX 78712 (USA)
Tel. +1 512 471 0194, Fax +1 512 471 9651, E-Mail H.Zakon@mail.utexas.edu

Definitions of and Tests for Convergence and Parallelism at the Molecular Level

At first glance, we might define convergent molecular evolution as the independent evolution of similar nucleotide or amino acid sequences in two unrelated molecules. Although this is one form that convergence can take, there are others. Doolittle [1994] defined four different types of convergent molecular evolution: functional, mechanistic, structural, and sequence convergence.

Molecules that serve the same function but have no sequence or structural similarity and carry out their function by entirely different mechanisms are said to be functionally convergent. For example, alcohol dehydrogenases in vertebrates and *Drosophila* bear no sequence similarity, their tertiary structures are entirely different, and they catalyze alcohol into acetaldehyde by different chemical reactions. Nevertheless, they are functionally similar in that they both remove a hydrogen from alcohol [Doolittle, 1994].

Mechanistic convergence occurs when the sequence and structure of molecules are very different but the mechanisms by which they act are similar. Serine proteases have evolved independently in bacteria (e.g., subtilisin) and vertebrates (e.g., trypsin). Despite their very different sequences and three-dimensional structures, they fold such that three amino acid residues – aspartic acid, serine, and histidine – form a binding pocket. These residues occur in very different positions in the linear sequence of the two proteins, and their formation of a binding pocket is not predictable by sequence alone. Recognizing mechanistic convergence is only possible once the three-dimensional structures of the proteins are known. Even this is just the starting point, as the critical function of the suspected residues must be probed with site-directed mutagenesis to be certain.

In some cases, molecules with very different amino acid sequences can assume similar structural motifs, and these may carry out similar functions. This is an example of structural convergence. For example, α helices and β sheets can be formed from a number of different amino acid sequences and are found in many proteins. A prime example given by Doolittle is of the remarkable similarity in fibronectin type III and immunoglobulin domains. These are composed of series of three and four stranded β sheets that are virtually identical in structure despite a lack of sequence similarity between these two molecules.

One caveat in identifying structural convergence is that it is not always easy to rule out that the molecules under consideration share a common ancestor. It is possible that their common ancestor possessed the same structure and that over time mutations accumulated in the two lineages that have altered the amino acid sequence completely, but spared the molecular structure [Doolittle, 1994]. The most convincing way to establish structural convergence is to estimate ancestral sequences of the two lineages, and then model or reconstruct the ancestral proteins. One could then determine whether the derived structural features emerged independently in the two lineages.

Finally, the type of convergent evolution that initially springs to mind is sequence convergence. Here, one or more critical amino acids or an amino acid sequence of two proteins come to resemble each other due to natural selection. This type of convergence is easiest to identify in molecules that are somewhat related, such as those within a gene superfamily; it is difficult to specify a particular amino acid site in a sequence otherwise. If the putative ancestral amino acids at a particular site were different in the ancestors of two proteins that now share an identical residue at that location, then convergent evolution may have occurred. If the inferred ancestral amino acids were the same in the ancestral proteins and independently changed to another amino acid in both lineages, then this may be a result of parallel evolution. The most frequently cited case of convergence and parallelism at the sequence level is the digestive enzyme lysozyme in a number of unrelated animals – the langur (a primate), the cow (an artiodactyl), and the hoatzin (a bird) – that have independently evolved the ability to utilize bacteria to digest cellulose [Kornegay, 1996; Zhang and Kumar, 1997]. Here a few specific residues have convergently evolved to allow digestion of cellulose-eating bacteria.

The assessment of sequence convergence would seem to be straightforward. However, we must first be convinced that the similarity between two sites or sequences is not due to common descent from a molecule with the same feature [Nei and Kumar, 2000]. This can be done by comparing sequences of a number of genes related to the molecules in question. These may be genes from different species or they could be gene families within a single species. This analysis is more informative when a larger number of genes are sampled, and it is critical that orthologs be used when comparisons are made among genes from a number of species. In other words, in gene families with many members, orthologs and paralogs should not be compared. If, for instance, a particular amino acid variant were to occur in all the members of a family of orthologous genes but not in their paralogs, a gene tree composed of both orthologs and paralogs might mistakenly suggest convergence where none existed, because some of the genes would possess that amino acid whereas others would not.

We must then be certain that any similarity is not due to chance. With only 20 amino acids, there is a probability of 1:20 that the same amino acid will occur at a given site in two molecules by chance. A string of 5 amino acids would be identical by chance alone with a probability of 1.25×10^{-4}, which is actually a high probability given the large number of amino acid sequences in the world. Rigorous statistical methods have been devised to differentiate convergence and parallelism from random similarity in amino acid sequences [Kornegay, 1996; Zhang and Kumar, 1997]. A further complication is that some amino acids substitute for others with little or no change of function (i.e., histidine for lysine) due to their charge or polarity.

Next, because convergence is the result of natural selection, the similarities must be important for the function of the molecule. We would assume that alterations in the amino acid(s) lead to non-functional or sub-optimally functioning molecules. As mentioned above, these kinds of questions can be addressed by site-directed mutagenesis and expression of altered proteins.

Finally, in making comparisons between two molecules, it is important to determine whether large-scale similarities between two seemingly distantly related genes are the result of horizontal gene transfer or recombination. For example, there is a region of about 250 residues in the C terminal of a calcium-activated K^+ channel (BK channels) that resembles a portion of a family of serine proteases. This is the region where calcium binds to the K^+ channel to activate it. Curiously, calcium binding to this region of some serine proteases alters their affinity for their substrates [Moss et al., 1996]. It is possible that this type of Ca^{2+}-activated K^+ channel arose from a recombination event involving a voltage-dependent K^+ channel and a serine protease. Large-scale recombination events such as this have an obvious potential to mislead in drawing conclusions about convergence.

Evolution of Opsins

Opsins are a family of proteins in the G-protein coupled receptor super-family that associate with a vitamin A-derived chromophore (retinal) and serve as the molecular transducer in all invertebrate and vertebrate photoreceptors. Retinal is photoisomerized upon absorption of light, but the amino acid composition of the opsin influences the absorption spectrum. There is variation in the number of opsins and their absorption maxima in the retinas of different species; five amino acid sites at different positions have been implicated in influencing the absorption spectrum of vertebrate opsins [Yokoyama and Radlwimmer, 2001]. This has raised the question of whether the same amino acids at these critical sites have convergently evolved to provide similar absorption spectra in different lineages.

One well-studied case is the evolution of trichromacy in Old and New World primates. Trichromacy is presumed to be adaptive for these species because the increased resolution of medium- and long-wavelengths (greens-yellows-reds) aids in the detection of fruits or edible young leaves in an arboreal habitat [Dominy and Lucas, 2001]. Most mammals, including diurnal prosimians, are thought to be dichromats. Thus, it has been suggested that trichromacy arose independently in New and Old World monkeys from dichromatic prosimians. Further support for this view comes from the different chromosomal organization of opsin genes in these two groups (fig. 1).

In New World monkeys, an opsin gene for short-wavelength (SW) sensitive cones (i.e., blue) is on an autosome, whereas there is *only a single opsin gene* on the X chromosome. The gene on the X chromosome has *three alleles.* These alleles encode opsins with different absorption maxima that span the mid- and long-wavelength (MW/LW) range. With this arrangement, all males are dichromats having the autosomal SW gene and *only one* X-linked MW or LW opsin allele. Females can be either dichromats or trichromats, depending on whether they inherit the same or different alleles on their X chromosomes [Nathans et al., 1986]. Old World monkeys, great apes and humans have the same autosomal SW cone pigment gene as New World monkeys, but have *separate* genes for MW and LW, both on the X chromosome. With this inheritance pattern, both sexes are trichromats.

A number of studies have used phylogenetic methods to infer how opsin genes evolved in primates and whether their amino acid sequences evolved convergently [Nei et al., 1997]. Using the sequence of the MW/LW gene from the dichromatic prosimian bushbaby *(Galago)* as an outgroup, comparisons of opsin genes in New and Old World primates and humans suggested that LW opsins evolved independently from MW opsins in both primate lineages along with parallel amino acid changes at some sites [Boissinot et al., 1998] (fig. 1A, B). Three critical residues in the MW opsin of both Old and New World monkeys are alanine, phenylalanine, and alanine, but at these same positions in LW opsins in both groups are serine, tyrosine and threonine. If we assume that LW opsins evolved from MW opsins, these three critical residues would have to have evolved in parallel.

However, a more recent study [Tan and Li, 1999] has shown a greater richness of opsin gene expression in prosimians than had been thought, which calls for a reinterpretation of this scenario. Sequencing fragments of opsin genes from a number of individuals of several families of prosimians has shown that some species have a LW gene, others have a MW opsin gene, and still others possess both as separate alleles of the same gene. In other words, some prosimians show the New World monkey pattern. This raises the possibility that all prosimians once possessed this pattern and that New World monkeys inherited the pattern from prosimian ancestors. If this is correct, the gene tree for opsins would have to be redrawn as in figure 1C.

But what might account for the different chromosomal organization of the opsin genes in New and Old World monkeys, a key point in the argument for convergent evolution? The most likely possibility is that an ancestral Old World monkey already had MW and LW opsin alleles that it inherited from its prosimian progenitors. These *alleles* subsequently came to be two separate *genes* on the same chromosome by a recombination event. This scenario was suggested by Dulai et al. [1994, 1999]. Curiously a similar genetic rearrangement may have occurred in a single group of New World primates, the howler monkeys, whose opsin gene numbers and chromosomal arrangements are more like those of Old World monkeys [Dulai et al., 1999].

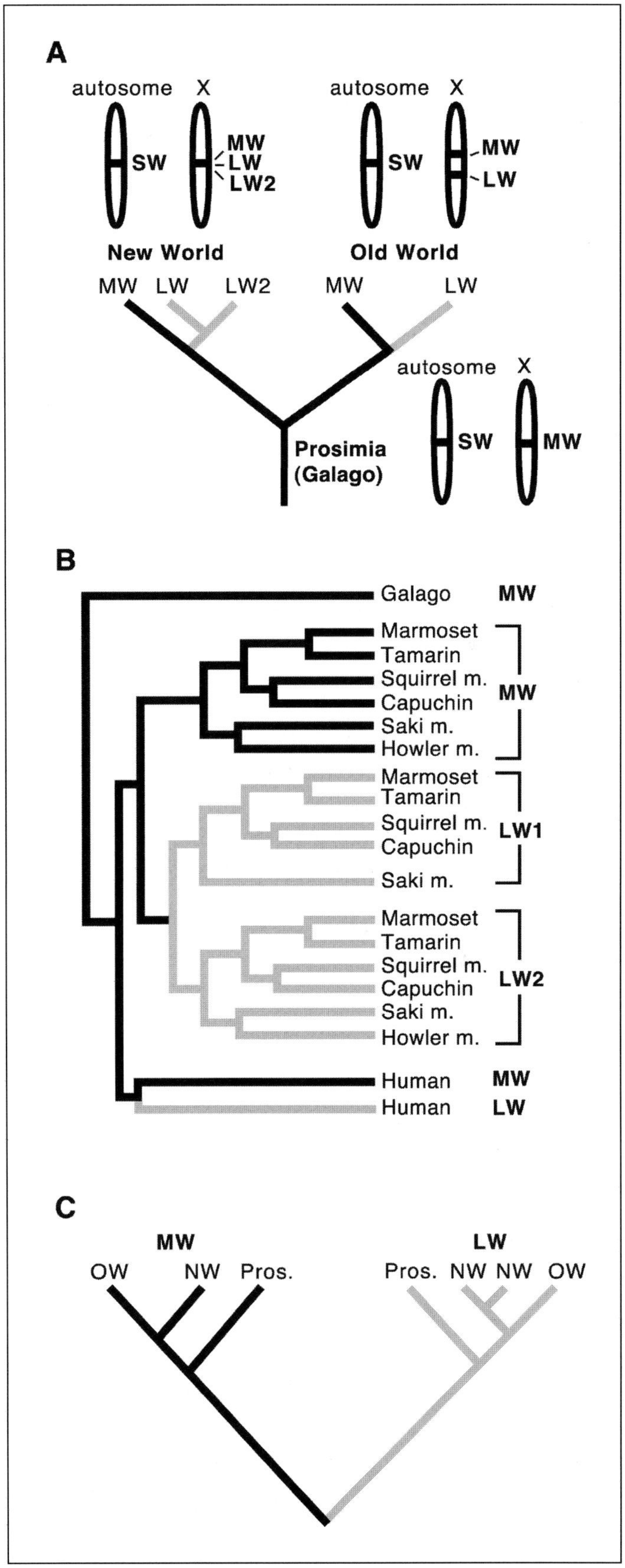

Fig. 1. Evolution of trichromacy in primates. **A** The genomic organization of short (SW), medium (MW) and long wavelength (LW) opsin genes in the genome of New World and Old World monkeys, and a representative prosimian, the bushbaby *(Galago)*. The LW and MW opsins of New World monkeys occur as three alleles of a single gene on the X chromosome, whereas they are different genes on the X chromosome in Old World monkeys, great apes, and humans. A schematic dendrogram illustrates the hypothesis that LW opsins evolved convergently in New and Old World lineages from a MW opsin gene that these groups shared in common with prosimians. **B** Gene tree of opsin sequences from New World primates and an Old World primate (human) rooted with a sequence from a prosimian, the bushbaby [redrawn from Boissinot et al., 1998]. **C** New schematic gene tree of MW and LW opsin genes incorporating the information that some prosimians (Pros.) have both MW and LW alleles at the opsin locus on the X chromosome, much as New World monkeys (NW) do. OW = Old World primates.

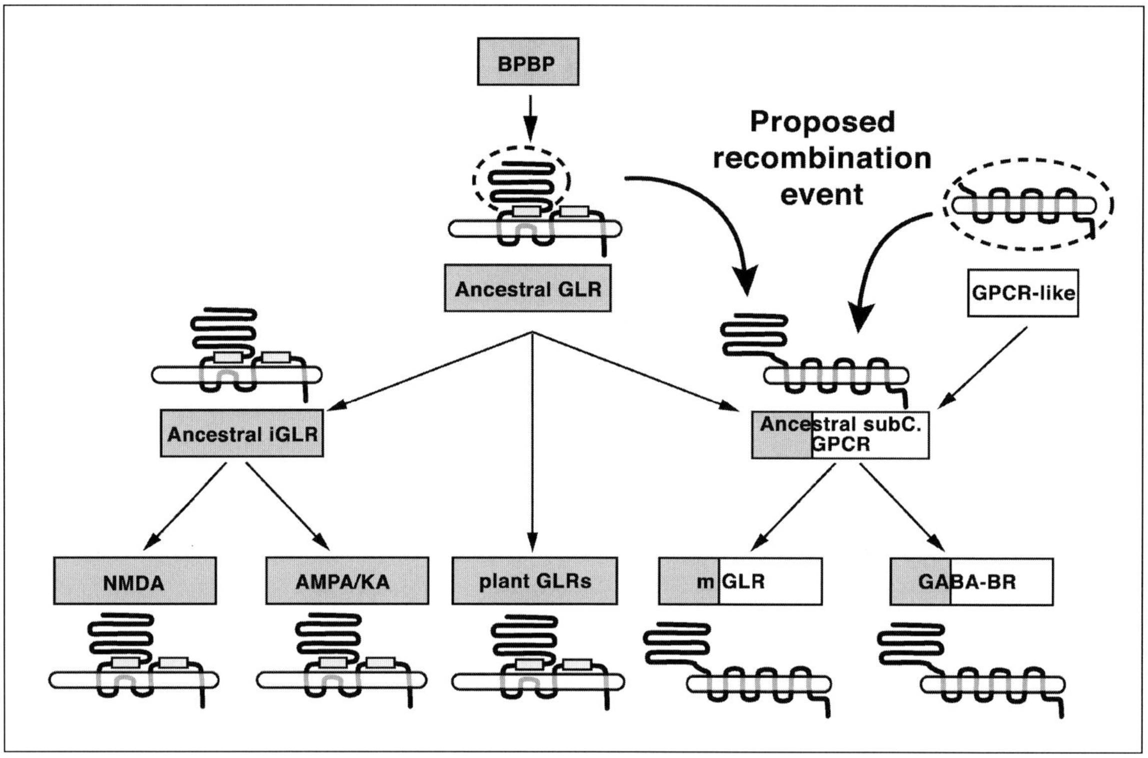

Fig. 2. Hypothesis for the evolution of the glutamate binding region of the ionotropic and metabotropic glutamate receptors. It has been proposed that the glutamate binding region of an ancestral ionotropic glutamate receptor came to lie upstream from a G protein-coupled receptor-like protein after a recombination event. Thus, the apparent similarity of these regions in ionotropic and metabotropic receptors is not due to convergence [redrawn from Turano et al., 2001]. AMPA/KA = α-amino-3-hydroxy-5-methyl-4-isoxazole propionate and kainate-type glutamate receptor; BPBP = bacterial periplasmic binding protein; GABA-BR = γ-amino-butyric acid class B receptor; GLR = glutamate receptor; GPCR = G-protein-coupled receptor; iGLR = ionotropic glutamate receptor; sub C GPCR = subfamily C of the GPCRs; mGLR = metabotropic glutamate receptor; NMDA = N-methyl-D-aspartate receptor.

What is instructive here is that without sufficient information about ancestral sequences we might be misled in our diagnosis of convergence. In addition, using chromosomal arrangements of genes to aid in identifying convergence is not foolproof.

Although the most likely scenario for the evolution of LW genes in primates argues against convergence, there is other evidence for convergent molecular evolution of opsins in vertebrates and invertebrates. It is believed that SW and MW/LW opsin genes diverged over 500 million years ago [Nei et al., 1997] but that further gene duplications that led to distinct MW and LW pigments in insects and vertebrates occurred independently. Comparisons of opsin gene sequences in butterflies, crayfish, and primates indicate that opsins with specific absorbance spectra are associated with particular amino acid substitutions at specific sites in these diverse phyla [Briscoe, 2000, 2001], and these could only have come about by convergent evolution.

Convergent Evolution of Neurotransmitter Receptors

Neurotransmitter receptors are functionally divided into two types: ionotropic, in which transmitter binding directly gates an ion channel, and metabotropic, in which the ligand indirectly opens an ion channel, typically with the mediation of a G protein. Sequence analysis has shown that the ionotropic and metabotropic receptors are unrelated, yet a number of neurotransmitters – glutamate, acetylcholine, GABA, serotonin – have receptors of both types. This raises questions about whether the sequence or structure of the ligand binding domains of the ionotropic and metabotropic receptors for a given neurotransmitter are similar and, if so, whether these similarities are due to convergence.

This question has been addressed recently for glutamate receptors. The AMPA, kainate, and NMDA glutamate receptors form a superfamily within the ligand-gated ion channels, whereas the metabotropic glutamate receptor (mGLR) and $GABA_B$ receptors are close relatives (subfamily C) within a superfamily of the large and functionally diverse group of G-protein coupled receptors.

Ionotropic glutamate receptor (iGLRs) genes have been found in plants; their sequence and structural similarity with those in animals suggest that they predate the plant-animal split [Lam et al., 1998; Chu et al., 1999] (fig. 2). Further, a motif similar to that of iGLRs has been recognized in the bacterial periplasmic binding protein (BPBP), a putative glutamate transporter, suggesting an even more ancient lineage for glutamate receptors [Stern-Bach et al., 1994]. Apparently, the ligand-binding domain of the mGLR is also similar to that of the BPBP [O'Hara et al., 1993]. This similarity suggests that both iGLRs and mGLRs might claim lineal descent from the bacterial BPBP (fig. 2).

However, a separate analysis of the N and C termini of the plant and animal GLRs and the $GABA_B$ R concluded that although the ligand binding domain (C terminus) of the mGLR is similar to the that of the iGLR, the N terminus of the *Arabidopsis* GLR is more similar to the N termini of the mGLR and the $GABA_B$ R than to that of the iGLR [Turano et al., 2001]. These authors suggested that a recombination event between the C terminus of an ancestral iGLR and a G protein coupled receptor gave rise to an ancestral chimeric gene from which the mGLR and $GABA_B$ R evolved (fig. 2).

This example emphasizes that one can be misled when making comparisons among long sequences and suggests that it is sometimes informative to compare smaller segments of large sequences. This also points out how recombination events can obscure true phylogenetic relationships even as they provide additional richness in evolutionary processes.

The story does not end there even for glutamate receptors, as there is still another group of receptors, the glutamate-activated anionic channels found in invertebrates [Cully et al., 1994] that are in yet another superfamily of neurotransmitter receptors. Did the glutamate binding domain of the glutamate-activated anionic channel also come about by recombination, or is this an example of convergence? These questions have yet to be addressed for this and other types of neurotransmitter receptors.

Convergent Evolution of Gap Junction Proteins in Vertebrates and Invertebrates

Gap junctions are areas of contact between the membranes of two cells that form an aqueous pore through which ions and small molecules pass freely. Gap junctions are made between neurons, glia, cardiac muscle, smooth muscle cells, and many other cell types [Lo, 1999]. A single gap junction is constructed from two hemichannels on each of two cells. A hemichannel is composed of proteins that form a hexameric array surrounding a pore in the membrane of one cell, and these then dock with a similar structure on another cell. These proteins, called connexins in vertebrates, form a gene family of over a dozen members. Connexins have four membrane spanning regions with two large extracellular loops and C and N termini inside the cell. The greatest conservation is found among the membrane spanning regions and the extracellular loops. The latter is not surprising, as this is the portion of the molecule responsible for contacting a hemichannel on a partner cell [Lo, 1999].

Gap junctions with generally similar properties have been described in invertebrates [Ganfornina et al., 1999]. Putative invertebrate gap junction protein genes have been cloned. These proteins, called innexins, appear to have identical structures of four membrane-spanning regions, conserved extracellular loops, and intracellular termini [Phelan et al., 1998]. However, there is no sequence homology between connexins and innexins on the amino acid or nucleotide level. The evidence for innexins being gap junction proteins is strong: mutations of innexin genes in *Drosophila* or *C. elegans* result in a defect in gap junctional coupling, injections of innexin anti-sense mRNA blocks gap junction formation, and expression of innexin genes in *Xenopus* oocytes promotes the formation of gap junctions between two injected oocytes [Phelan et al., 1998].

Is this a case of structural convergent evolution or, rather, a case of shared ancestry masked by the gradual accumulation of amino acid mutations that preserve the secondary structure of the protein? This question could be best resolved by obtaining the sequence of gap junction proteins in a number of pivotal organisms, such as jellyfish and amphioxus. Although this information is currently unavailable, innexins have been cloned in two annelids (a leech and a polychaete worm), and they show clear

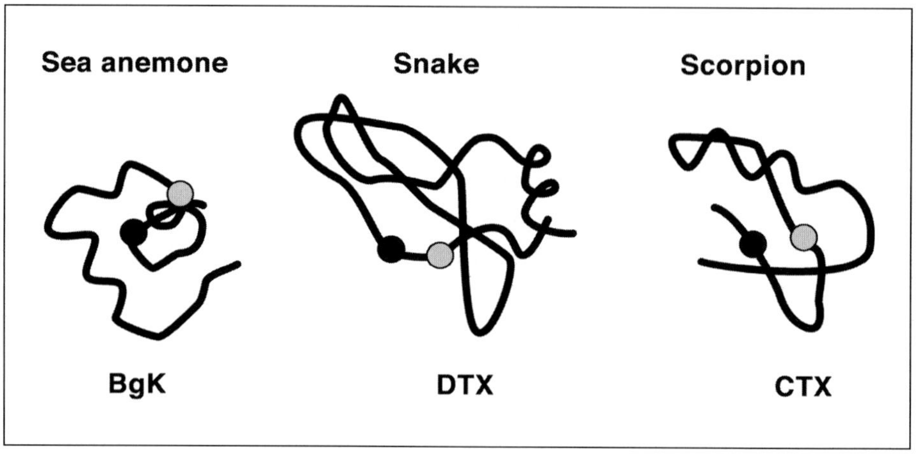

Fig. 3. Convergent evolution at two critical amino acid sites in K⁺ channel blocking venoms from three different species. The filled circle represents a hydrophobic residue and the open circle represents a lysine. Despite the fact that these residues are located in different places in the peptide sequence of the different venoms, they are near each other (6–7 Å) in the folded molecules [redrawn from Gasparini et al., 1998]. This is an example of mechanistic convergence according to Doolittle [1994]. BgK = a K⁺ channel blocking peptide from the sea anemone *Bunodosoma granulifera*; DTX = dendrotoxin; CTX = charybdotoxin.

homology with the innexins of other invertebrates and none to connexins [Dykes et al., 2001; Potenza et al., 2002].

Convergent Evolution of Neurotoxins

A number of different taxa, including sea anemones (cnidaria), cone snails (molluscs), scorpions (arthropods), and snakes (vertebrates), synthesize ion channel-blocking peptide toxins. The biology of venoms is rich with examples of divergence and convergence. These animals make toxins to a number of ion channels and ligand-gated receptors, some of which also show convergent evolution of amino acid residues [such as those against nicotinic acetylcholine receptors; Dufton et al., 1989; Servent et al., 1998], but this review will focus on K⁺ channels as a clear example.

African mamba snakes produce a number of Kv-type K⁺ channel-blocking toxins, generally referred to as dendrotoxins. The dendrotoxins are related to a group of proteins called the Kunitz serine protease inhibitors; in fact, some dendrotoxins still retain weak trypsin inhibitory activity [Harvey, 2001; Sigle et al., 2001]. Peptides with K⁺ channel blocking properties have also been isolated from sea anemones [Schweitz et al., 1995]. One group of peptides, the kalicludines, bear sequence homology with the dendrotoxins and the serine protease inhibitors, and even function as trypsin-inhibitors. This appears to be a case of parallelism in which two groups of neurotoxins with similar functions evolved independently from a common class of ancestral proteins, the serine protease inhibitors. The other alternative, that similar neurotoxins were made in all relatives of sea anemones and snakes (i.e., most animals) and lost in all lineages but these two, seems untenable.

Sea anemones produce another molecularly unrelated group of toxins, including kaliseptine and BgK toxins. Interestingly, structural analysis of the toxins and mutagenesis of potentially important residues of the toxins have shown that the snake toxin α-dendrotoxin, the anemone toxin BgK, and the scorpion toxin charybdotoxin all possess certain key residues – a lysine and a nearby hydrophobic residue (Tyr, Phe, Leu) – in common, despite their completely different structure and folding, and their independent origins [Dauplais et al., 1997; Gasparini et al., 1998; Alessandri-Haber et al., 1999] (fig. 3). The cone snail K⁺ channel blocker κ conotoxin PVIIA competes against dendrotoxin for the same site on the Kv channel; although it has not yet been subjected to analysis by mutagenesis and substitution of amino acids, it also possesses a lysine close to a phenylalanine, suggesting that it too might block the channel by the same mechanism [Savarin et al., 1998].

This example of mechanistic convergent evolution of similar residues in four distinct molecules may not be surprising, as there must be a limited number of solutions to achieving a high affinity blockade of Kv-type K⁺ [Tytgat et al., 1995], and the fundamental structure of the K⁺ channel pore is more or less the same from bacteria to eukaryotes [Doyle et al., 1998; Sokolova et al., 2001]. In each case, the positively-charged lysine is thought to lodge in the mouth of the pore, presumably near a K⁺ binding site. The role of the hydrophobic residue is not known.

Perhaps more striking than the convergent evolution of a few critical residues in otherwise distinct peptides, is the convergent evolution of the fundamental 'strategy' for the production of large batteries of potent ion-channel blocking toxins. Cone snails are notable for the staggeringly large number of toxins that they produce; it is conservatively estimated that some 500 species of cone snails produce ~50,000 distinct toxins, with each species capable of making 100–300 [Olivera and Cruz, 2001]. There are 5–10 conotoxin gene superfamilies, each of which is represented by hundreds of different genes. Recent studies on the molecular organization of the conotoxin genes have illustrated the mechanisms responsible for that abundance of toxins.

Mature conotoxin peptides are 10–35 amino acids long. Like other peptides, the newly translated peptide also has a pre- (or signal sequence) pro-region at the N terminal. Conotoxin genes are organized with one intron between the pre- and pro-regions, and another between the pro-region and the region that codes for the mature peptide (which is a single exon). Comparisons between these portions of the genes show a striking result: whereas introns tend to be far less conserved than exons in most genes, the sequence of the exons encoding the mature peptide show much greater variation across species than the pre-pro-region and the introns [Duda and Palumbi, 1999; Conticello et al., 2001; Espiritu et al., 2001]. This diversity stems from gene duplication and some amount of recombination between the exons of different genes, common mechanisms for generating diversity. However, beyond that, the region encoding the mature peptide seems to be hypermutable. Perhaps these genes recruit an unusual DNA polymerase somewhat like a DNA polymerase V. This polymerase, which is expressed in bacteria under stress conditions, makes frequent mistakes, thereby rapidly introducing genetic diversity into the population [Conticello et al., 2001].

Scorpions also possess a large number of neurotoxins and these may well have evolved by a similar process of extensive gene duplication and hyper-mutation [Froy et al., 1999]. Despite the difference in function (blockers of Na^+, Ca^{2+}, Cl^- and K^+ channels), size, and structure, all scorpion neurotoxin genes share a single intron that interrupts a codon in the signal sequence. The similarity in intron position and the rarity of an intron dividing a codon strongly suggest that these toxins evolved from a common ancestor. Thus, the processes of large-scale duplication and hypermutation may have evolved convergently in these two taxa.

It is not yet known if snakes have evolved a similar strategy in the evolution of dendrotoxins, but it is suggested that such a strategy evolved in the production of phospholipase A_2 – and some serine protease-based venoms [Deshimaru et al., 1996; Kini and Chan, 1999].

Evolution of Toxin Resistance

Tetrodotoxin is a potent non-peptide neurotoxin that blocks many types of Na^+ channels with nanomolar affinity. Although it is most commonly associated with pufferfish of the genus *Fugu*, it is found in a number of other groups as well, including other fish, newts, the ghost crab, blue-ringed octopus, and some poison arrow frogs. A clue to this surprising phylogenetic diversity is that TTX is likely to be of bacterial origin [Yotsu et al., 1987]. Thus, symbiosis between TTX-producing bacteria and animals might have evolved multiple times.

Many animals that make toxins sequester them in glands so that they are not exposed to their own toxins. TTX, however, is distributed in various body tissues of those organisms possessing it. For example, TTX-producing bacteria have been cultured from the skin and the gut in pufferfish [Yotsu et al., 1987], yet TTX is in the highest concentration in the ovary and liver. The fact that TTX is in general circulation in the body means that the Na^+ channels of the host organism are exposed and must somehow become immune to its toxicity. Exposure to TTX in concentrations manyfold higher than the lethal dose for mice or other teleost fish has no effect on pufferfish [Kidokoro et al., 1974; Saito et al., 1984] and the binding affinity of pufferfish muscle and brain for TTX is extraordinarily low, in the millimolar range [Yotsu-Yamashita et al., 2000]. Similarly, in the newt *(Notophthalmus viridescens)*, TTX has little physiological effect [Kaneko et al., 1997]. This suggests that the Na^+ channel genes in pufferfish and newts, and perhaps other TTX-bearing organisms, have convergently evolved to make a Na^+ channel that is insensitive to TTX.

Evolution of TTX insensitivity in pufferfish and newts can be 'modeled' by the rapid evolution of insecticide (pyrethroid) insensitivity by a few specific point mutations of the Na^+ channel gene in natural populations of flies, cockroaches, and other insects within a few decades of insecticide use [ffrench-Constant et al., 1998]. It is easy to see how spontaneous mutations that conferred toxin resistance without altering normal function would be highly selected. However, the story in vertebrates is more complicated and interesting. This is because, unlike insects that have a single Na^+ channel gene, vertebrates possess multiple Na^+ channel genes. The total number of these genes has not been determined yet for any TTX-producing species; however, teleost fish have at least 6 distinct Na^+ channels [Lopreato et al., 2001], and amphibians likely have that number or more. Thus, TTX-insensitivity most likely evolved in multiple paralogous genes within an organism.

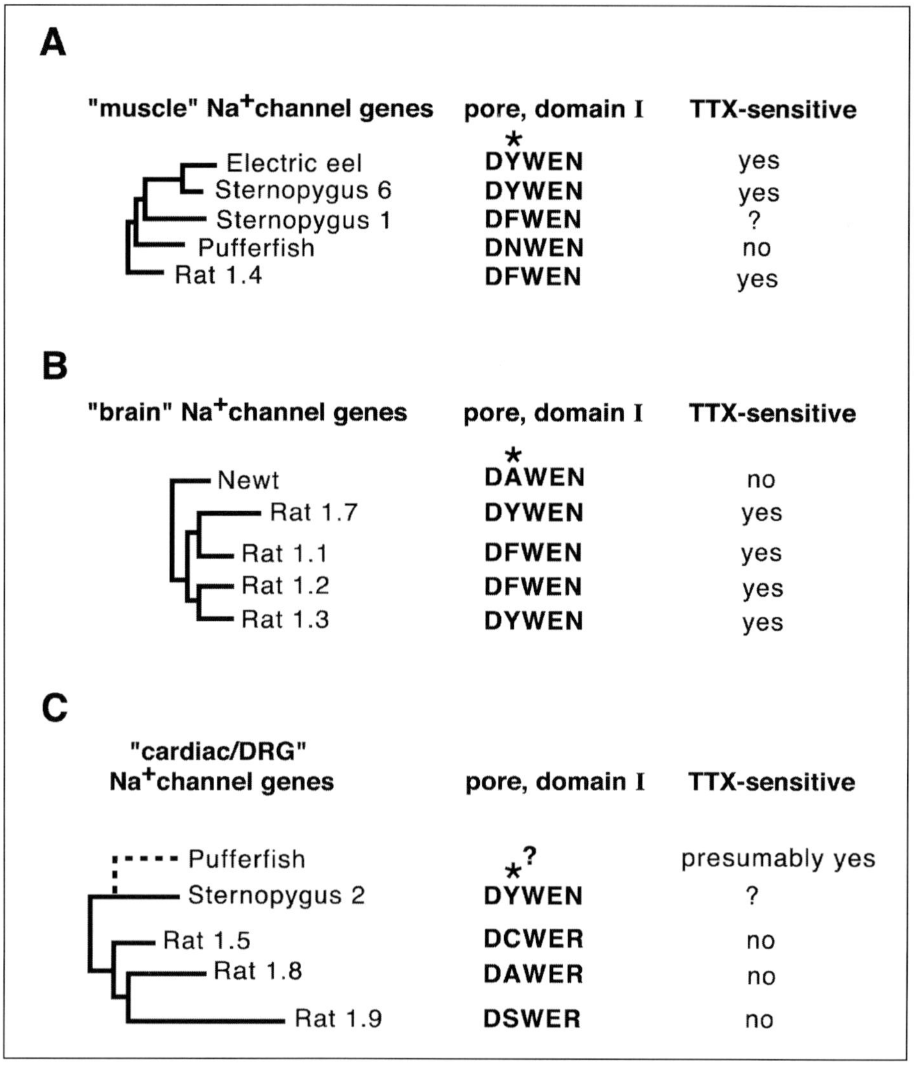

Fig. 4. Amino acid sequences in the pore of the Na$^+$ channel in TTX-producing species (pufferfish, newt) and non toxin-producing species. **A** Sequence of Na$^+$ channel in muscle (rat 1.4, *Sternopygus* 1) or the muscle-derived electric organ (eel and *Sternopygus* 6). All have Y (tyrosine) or F (phenylalanine), which are aromatic amino acids, whereas the pufferfish channel has N (arginine) which is a non-aromatic amino acid. (This site is indicated by an asterisk in all parts of this figure.) This substitution hinders the binding of TTX to the pufferfish's Na$^+$ channel. **B** In a Na$^+$ channel expressed in the brain, the TTX-producing newt has the non-aromatic amino acid A (alanine) at that site. **C** Three genes expressed in the heart (rat 1.5) or in dorsal root ganglia (DRG) in mammals (rat 1.8, 1.9) contain the amino acids C, A, or S (cysteine, alanine, or serine) which confer TTX-resistance. The cardiac Na$^+$ channel gene has not yet been cloned in pufferfish; however, there is a Y in that position in the weakly electric fish *Sternopygus*. Without sequences from more species, the evolution of this site is ambiguous, but it is likely that the ancestral sequence was Y and that it may have mutated to a non-aromatic amino acid independently in mammals and in pufferfish.

The Na$^+$ channel is a single large protein with four repeating domains. Each of these domains has a highly conserved region that contributes to the formation of the pore. A comparison of the amino acid sequence of the pore regions of Na$^+$ channels in TTX-sensitive and insensitive organisms reveals some striking differences (fig. 4). In the first domain of the muscle Na$^+$ channel in rat or another teleost (the weakly electric fish, *Sternopygus macrurus*) is an amino acid sequence D(Y/F)WEN (or aspartate-tyrosine or phenylalanine-tryptophan-glutamate-asparagine). In some genes there is a Y in others an F, but both are aromatic amino acids. In the same position in the pufferfish muscle Na$^+$ channel gene, is an N, which is a non-aromatic amino acid. A comparison with this same domain in a brain Na$^+$ channel gene in rat and the newt shows D(Y/F)WEN in the rat sequence and an A (alanine, another non-aromatic amino acid) in the newt [Kaneko et al., 1997]. This suggests that this critical amino acid was mutated and this mutation was conserved to decrease TTX binding affinity. Indeed, mutagenesis of this residue in the mammalian muscle Na$^+$ channel gene confers on the Na$^+$ channel a much lower affinity for TTX. This strongly supports the idea that this particular site with a highly conserved aromatic amino acid underwent parallel evolution to become a non-aromatic amino acid to decrease TTX binding.

Has this happened in the other Na$^+$ channel genes in these organisms? Three of the ten Na$^+$ channel genes in mammals are TTX-insensitive. These three genes reside on the same chromosome and likely evolved as duplications of an ancestral TTX-insensitive gene. If this gene were present in all vertebrates (or at least in teleosts and tetrapods) and this were the ancestral sequence of this gene in all vertebrates, there would have been no need for any further evolution of this gene to confer TTX resistance. A phylogenetic analysis of Na^+ channel genes in a teleost and mammals supports the idea that this gene had an ancestor that predated the teleost-tetrapod split [Lopreato et al., 2001]. However, because the sequence of this gene is DFWEN in another teleost, it suggests either that the ancestral gene was not TTX-resistant and that TTX-resistance evolved in parallel in pufferfish and mammals (and probably other tetrapods) or that the ancestral gene was TTX-insensitive and that this was lost in the electric fish. This question can be easily resolved when sequences are available for other species.

If we assume for the moment that the ancestral condition was for an aromatic amino acid at this site, then we must ask why the amino acid changed to a non-aromatic amino acid in mammals. The most likely possibility is that a change to a non-aromatic amino acid at this site was selected for some other reason; the most plausible is that the rates of channel opening and closing (the activation and inactivation kinetics) of channels with a non-aromatic amino acid at this site in the pore are slow. The action potentials from the cells that express these sodium channel genes in mammals are long duration. Thus, it is possible that it is coincidental that these genes are TTX-insensitive in mammals.

Finally, the evolution of neurotoxins for prey capture or, conversely, deterrence of predation puts a strong selection pressure on the animals with which the toxin-producing species routinely interact. The California newt *(Taricha torosa)* is a major prey item for the garter snake *(Thamnophis sirtalis)*. Interestingly, where the garter snake and newt populations overlap, garter snakes have evolved the ability to eat the newts with no ill effects, whereas populations of the same species of snake that do not feed on newts in the wild become paralyzed when they are fed newts in the laboratory. It has been suggested that garter snakes have convergently evolved TTX-insensitive Na^+ channels [Motychak et al., 1999]. Once the genes for Na^+ channels in garter snakes are cloned, it will be interesting to see if they show convergent or parallel substitutions at the same sites as in newts and pufferfish.

Conclusion

Convergence occurs at the molecular level but in different forms, as outlined by Doolittle [1994]. Structural convergence may be the most difficult to prove with certainty because similar structures can be attained with many different amino acid sequences, and similar structural motifs appear in so many proteins. The proof of mechanistic convergence depends on the establishment of three-dimensional structure followed by functional analysis of the suspected residues. This can be achieved for peptides and small proteins, and a number of interesting cases of mechanistic convergence are known (substrate binding sites in serine proteases, pore-binding sites in K^+ channel blockers). Convergence at the sequence level is easily studied and statistical tests exist for distinguishing true sequence convergence from random changes in sequence. Again, a number of examples of sequence convergence are documented; these are usually of one or a few amino acids that play key functional roles, such as amino acids influencing absorbance in opsins or toxin binding in Na^+ channels.

In the examples of convergence described above, the similarities are in the finished product – the protein. Convergence might also occur in the non-coding regions of genes. Certain motifs appear in the transcriptional control regions of diverse genes, but when is this due to common descent, recombination, or convergence? Phylogenetic analyses of non-coding regions are in their infancy as large-scale comparisons of these regions are difficult. Comparisons of 'upstream' control regions between similar genes in related organisms (i.e., different species of *Drosophila*; mouse vs. human) are underway [Bohm et al., 2000].

A variation of this theme is studying convergence of molecular mechanisms in highly specialized systems at the genetic level. So, for example, the strategy of 'toxin production by combinatorial chemistry' observed in scorpions, snakes, and cone snails shows convergence on a number of features of gene organization for enhancement of gene duplication and recombination, and for the induction of hypervariability. One wonders whether other gene families with large numbers, that underwent explosive bouts of duplications, and with rich species diversity such as mammalian and, perhaps insect, olfactory receptors, may have convergently evolved similar mechanisms [Sosinsky et al., 2000; Glusman et al., 2001; Störtkuhl and Kettler, 2001].

Acknowledgements

I would like to thank Gwen Gage for artwork, and NIH for funding.

References

Alessandri-Haber, N., A. Lecoq, S. Gasparini, G. Grangier-Macmath, G. Jacquet, A. Harvey, C. de Medeiros, E. Rowan, M. Gola, A. Ménez, and M. Crest (1999) Mapping the functional anatomy of BgK on Kv1.1, Kv1.2, and Kv1.3. Clues to design analogs with enhanced selectivity. J. Biol. Chem., *274:* 35653–35661.

Bohm, R., B. Wang, R. Brenner, and N. Atkinson (2000) Transcriptional control of Ca^{2+}-activated K^+ channel expression: Identification of a second, evolutionarily conserved, neuronal promoter. J. Exp. Biol., *203:* 693–704.

Boissinot, S., Y. Tan, S.-K. Shyue, H. Schneider, I. Sampaio, K. Neiswanger, D. Hewett-Emmett, and W.-H. Li (1998) Origins and antiquity of X-linked triallelic color vision systems in New World monkeys. Proc. Natl. Acad. Sci. USA, *95:* 13749–13754.

Briscoe, A. (2000) Six opsins from the butterfly *Papillio glaucus*: Molecular phylogenetic evidence for paralogous origins of red-sensitive visual pigments in insects. J. Mol. Evol., *51:* 110–121.

Briscoe, A. (2001) Functional diversification of lepidopteran opsins following gene duplication. Mol. Biol. Evol., *18:* 2270–2279.

Chu, J., R. DeSalle, H.-M. Lam, L. Meisel, and G. Coruzzi (1999) Molecular evolution of glutamate receptors: A primitive signalling mechanism that existed before plants and animals diverged. Mol. Biol. Evol., *16:* 826–838.

Conticello, S., Y. Gilad, N. Avidan, E. Ben-Asher, Z. Levy, and M. Fainzilber (2001) Mechanisms for evolving hypervariability: The case of conopeptides. Mol. Biol. Evol., *18:* 120–131.

Cully, D., D. Vassilatis, K. Liu, P. Paress, L. Van der Ploeg, J. Schaeffer, and J. Arena (1994) Cloning of an avermectin-sensitive glutamate-gated chloride channel from *Caenorhabditis elegans.* Nature, *371:* 707–711.

Dauplais, M., A. Lecoq, J. Song, J. Cotton, N. Jamin, B. Gilquin, C. Roumestand, C. Vita, C. de Medeiros, E. Rowan, A. Harvey, and A. Ménez (1997) On the convergent evolution of animal toxins. Conservation of a diad of functional residues in potassium channel-blocking toxins with unrelated structures. J. Biol. Chem., *272:* 4302–4309.

Deshimaru, M., T. Ogawa, K. Nakashima, I. Nobuhisa, T. Chijiwa, Y. Shimohigashi, Y. Fukumaki, M. Niwa, I. Yamashina, S. Hattori, and M. Ohno (1996) Accelerated evolution of crotalinae snake venom gland serine proteases. FEBS Letters, *397:* 83–88.

Dominy, N., and P. Lucas (2001) Ecological importance of trichromatic vision to primates. Nature, *410:* 363–366.

Doolittle, R. (1994) Convergent evolution: The need to be explicit. Trends Biochem. Sci., *19:* 15–18.

Doyle, D.A., J. Morais Cabral, R.A. Pfuetzner, A. Kuo, J.M. Gulbis, S.L. Cohen, B.T. Chait, and R. MacKinnon (1998) The structure of the potassium channel: Molecular basis of K^+ conduction and selectivity. Science, *280:* 69–77.

Duda, T.J., and S. Palumbi (1999) Molecular genetics of ecological diversification: Duplication and rapid evolution of toxin genes of the venomous gastropod *Conus*. Proc. Natl. Acad. Sci. USA, *96:* 6820–6823.

Dufton, M.J., P. Bladon, and A.L. Harvey (1989) Identification of a locality in snake venom alpha-neurotoxins with a significant compositional similarity to marine snail alpha-conotoxins: Implications for evolution and structure/activity. J. Mol. Evol., *29:* 355–366.

Dulai, K.S., J.K. Bowmaker, J.D. Mollon, and D.M. Hunt (1994) Sequence divergence, polymorphism and evolution of the middle-wave and long-wave visual pigment genes of great apes and Old World monkeys. Vis. Sci., *34:* 2483–2491.

Dulai, K.S., M. von Dornum, J.D. Mollon, and D.M. Hunt (1999) The evolution of trichromatic color vision by opsin gene duplication in New World and Old World Primates. Genome Res., *9:* 629–638.

Dykes, I., J. Bacon, and J. Davies (2001) Cloning and characterization of a leech innexin. Soc. Neurosci. Abstr., *27:* 813.5.

Espiritu, D., M. Watkins, V. Dia-Monje, G. Cartier, L. Cruz, and B. Olivera (2001) Venomous cone snails: Molecular phylogeny and the generation of toxin diversity. Toxicon, *39:* 1899–1916.

ffrench-Constant, R., B. Pittendrigh, A. Vaughan, and N. Anthony (1998) Why are there so few resistance associated mutations in insecticide target genes? Phil. Trans. Roy. Soc. Lond. B, *353:* 1685–1693.

Froy, O., T. Sagiv, M. Poreh, D. Urbach, N. Zilberberg, and M. Gurevitz (1999) Dynamic diversification from a putative common ancestor of scorpion toxins affecting sodium, potassium, and chloride channels. J. Mol. Evol., *48:* 187–196.

Ganfornina, M., D. Sánchez, M. Herrera, and M. Bastiani (1999) Developmental expression and molecular characterization of two gap junction channel proteins expressed during embryogenesis in the grasshopper *Schistocerca americana.* Dev. Gen., *24:* 137–150.

Gasparini, S., J. Danse, A. Lecoq, S. Pinkasfeld, S. Zinn-Justin, L. Young, C. de Medeiros, E. Rowan, A. Harvey, and A. Ménez (1998) Delineation of the functional site of α-dendrotoxin. The functional topographies of dendrotoxins are different but share a conserved core with those of other Kv1 potassium channel-blocking toxins. J. Biol. Chem., *273:* 25393–25403.

Glusman, G., I. Yanai, I. Rubin, and D. Lancet (2001) The complete human olfactory subgenome. Genome Res., *11:* 685–702.

Harvey, A. (2001) Twenty years of dendrotoxins. Toxicon, *39:* 15–26.

Kaneko, Y., G. Matsumoto, and Y. Hanyu (1997) TTX resistivity of Na^+ channel in newt retinal neuron. Biochem. Biophys. Res. Comm., *240:* 651–656.

Kidokoro, Y., A. Grinnell, and D. Eaton (1974) Tetrodotoxin sensitivity of muscle action potentials in pufferfishes and related fishes. J. Comp. Physiol., *89:* 59–72.

Kini, R., and Y. Chan (1999) Accelerated evolution and molecular surface of venom phospholipase A_2 enzymes. J. Mol. Evol., *48:* 125–132.

Kornegay, J. (1996) Molecular genetics and evolution of stomach and nonstomach lysozymes in the hoatzin. J. Mol. Evol., *42:* 676–684.

Lam, H.-M., J. Chiu, M.-H. Hsieh, L. Meisel, I. Oliveira, M. Shin, and G. Coruzzi (1998) Glutamate-receptor genes in plants. Nature, *396:* 125–126.

Lo, C. (1999) Genes, gene knockouts, and mutants in the analysis of gap junctions. Dev. Gen., *24:* 1–4.

Lopreato, G., Y. Lu, A. Southwell, N. Atkinson, D. Hillis, T. Wilcox, and H. Zakon (2001) Evolution and divergence of sodium channel genes in vertebrates. Proc. Natl. Acad. Sci. USA, *98:* 7588–7592.

Moss, G., J. Marshall, M. Morabito, J. Howe, and E. Moczydlowski (1996) An evolutionarily conserved binding site for serine proteinase inhibitors in large conductance calcium-activated potassium channels. Biochemistry, *35:* 16024–16035.

Motychak, J., E. Brodie III, and E.D.J. Brodie (1999) Evolutionary response of predators to dangerous prey: Preadaptation and the evolution of tetrodotoxin resistance in garter snakes. Evolution, *53:* 1528–1535.

Nathans, J., D. Thomas, and D. Hogness (1986) Molecular genetics of human color vision: The genes encoding blue, green, and red pigments. Science, *232:* 193–203.

Nei, M., and S. Kumar (2000) Molecular Evolution and Phylogenetics. Oxford University Press, Oxford.

Nei, M., J. Zhang, and S. Yokoyama (1997) Color vision of ancestral organisms of higher primates. Mol. Biol. Evol., *14:* 611–618.

O'Hara, P., P. Sheppard, H. Thogerson, D. Venezia, B. Haldeman, V. McGrane, K. Houame, C. Thomsen, T. Gilbert, and E. Mulvihill (1993) The ligand binding domain in metabotropic glutamate receptors is related to bacterial periplasmic binding proteins. Neuron, *11:* 41–52.

Olivera, B., and L. Cruz (2001) Conotoxins, in retrospect. Toxicon, *39:* 7–14.

Phelan, P., J. Bacon, J. Davies, L. Stebbings, M. Todman, L. Avery, R. Baines, T. Barnes, C. Ford, S. Hekimi, R. Lee, J. Shaw, T. Starich, K. Curtin, Y.-A. Sun, and R. Wyman (1998) Innexins: A family of invertebrate gap-junction proteins. Trends Gen, *14:* 348–349.

Potenza, L., R. del Gaudio, L. Rivieccio, G.M. Russo, and G. Geraci (2002) Cloning and molecular characterization of the first innexin of the phylum Annelida – Expression of the gene during development. J. Mol. Evol., *54:* 312–321.

Pough, F., J. Heiser, and W. McFarland (1996) Vertebrate Life. Prentice Hall, Upper Saddle Valley, NJ.

Saito, T., J. Maruyama, S. Kanoh, J.-K. Jeon, T. Noguchi, T. Harada, O. Murata, and K. Hashimoto (1984) Toxicity of the cultured pufferfish *Fugu rubripes rubripes* along with their resistibility against tetrodotoxin. Bull. Jpn Soc. Sci. Fish., *50:* 1573–1575.

Savarin, P., M. Guenneugues, B. Gilquin, H. Lamthanh, S. Gasparini, S. Zinn-Justin, and A. Ménez (1998) Three-dimensional structure of κ-conotoxin PVIIA, a novel potassium channel-blocking toxin from cone snails. Biochemistry, *37:* 5407–5416.

Schweitz, H., T. Bruhn, E. Guillemare, D. Moinier, J.-M. Lancelin, L. Béress, and M. Lazdunski (1995) Kalicludines and kaliseptine. J. Biol. Chem., *270:* 25121–25126.

Servent, D., H.-L. Thanh, S. Antil, D. Bertrand, P.-J. Corringer, J.-P. Changeux, and A. Ménez (1998) Functional determinants by which snake and cone snail toxins block the a 7 neuronal nicotinic acetylcholine receptors. J. Physiol., *92:* 107–111.

Sigle, R., M. Hackett, and S. Aird (2001) Primary structures of four trypsin inhibitor E homologs from venom of *Dendroaspis angusticeps*: Structure-function comparisons with other dendrotoxin homologs. Toxicon, *40:* 297–308.

Sokolova, O., L. Kolmakova-Partensky, and G.N. Grigorieff (2001) Three-dimensional structure of a voltage-gated potassium channel at 2.5 nm resolution. Structure, *9:* 215–220.

Sosinsky, A., G. Glusman, and D. Lancet (2000) The genomic structure of human olfactory receptor genes. Genomics, *70:* 49–61.

Stern-Bach, Y., B. Bettler, M. Hartley, P. Sheppard, P. O'Hara, and S. Heinemann (1994) Agonist selectivity of glutamate receptors is specified by two domains structurally related to bacterial amino acid-binding proteins. Neuron, *13:* 1345–1357.

Störtkuhl, K., and R. Kettler (2001) Functional analysis of an olfactory receptor in *Drosophila melanogaster*. Proc. Natl. Acad. Sci. USA, *98:* 9381–9385.

Tan, Y., and W.-H. Li (1999) Trichromatic vision in prosimians. Nature, *402:* 36.

Turano, F., G. Panta, M. Allard, and P. van Berkum (2001) The putative glutamate receptors from plants are related to two superfamilies of animal neurotransmitter receptors via distinct evolutionary mechanisms. Mol. Biol. Evol., *18:* 1417–1420.

Tytgat, J., T. Debon, E. Carmeliet, and P. Daenens (1995) The α-dendrotoxin footprint on a mammalian potassium channel. J. Biol. Chem., *270:* 24776–24781.

Yokoyama, S., and F. Radlwimmer (2001) The molecular genetics and evolution of red and green color vision in vertebrates. Genetics, *158:* 1697–1710.

Yotsu, M., T. Yamazaki, Y. Meguro, A. Endo, M. Murata, H. Naoki, and T. Yasumoto (1987) Production of tetrodotoxin and its derivatives by *Pseudomonas* sp. isolated from the skin of a pufferfish. Toxicon, *25:* 225–228.

Yotsu-Yamashita, M., K. Nishimori, Y. Nitanai, M. Isemura, A. Sugimoto, and T. Yasumoto (2000) Binding properties of ^{3}H-PbTx-3 and ^{3}H-saxitoxin to brain membranes and to skeletal muscle membranes of puffer fish *Fugu pardalis* and the primary structure of a voltage-gated Na^+ channel a-subunit (fMNa1) from skeletal muscle of *F. pardalis*. Biochem. Biophys. Res. Commun., *267:* 403–412.

Zhang, J., and S. Kumar (1997) Detection of convergent and parallel evolution at the amino acid sequence level. Mol. Biol. Evol., *14:* 527–536.

Brain, Behavior and Evolution

Brain Behav Evol 2002;59:262–272

Convergences in the Modular and Areal Organization of the Forebrain of Mammals: Implications for the Reconstruction of Forebrain Evolution

Jon Kaas

Department of Psychology, Vanderbilt University, Nashville, Tenn., USA

Key Words

Cerebral cortex · Thalamus · Evolution · Primates · Cortex · Visual system · Superior colliculus · Tree shrews · Squirrels

Abstract

Early efforts to reconstruct the course of the evolution of the human brain relied on comparing the brains of a few related mammals with brains at successively higher levels of complexity. This Clark or ladder of levels approach is now seen as having limited usefulness in that species are not easily assigned to levels, and extant mammals are now recognized as mosaics of primitive and derived features. In addition, direction of change does not necessarily proceed from simple to complex, small to large, or diffuse to differentiated. A modern cladistic approach reconstructs the brains of ancestors by identifying brain characters within and across phylogenetic groups (clades), and uses parsimony or likelihood to infer direction of change and distinguish ancestral features from independently evolved convergences. Unfortunately, an idealized cladistic approach is often difficult to realize because characters may be hard to identify and validate, key species may be unavailable for study, and broadly based comparative studies can be costly, poorly funded, and labor intensive. Thus, many investigators pursue a truncated approach that is superficially Clark-like but conceptually cladistic. A truncated approach that relies on the extensive study of a few species may compensate for weaknesses by including niche-matched species that offer the opportunity to estimate the likelihood of similar brain features evolving as convergent adaptations. Because inferences about the brains of the primate ancestor are often made from the brains of tree shrews, we compare the brains of squirrel-like tree shrews with the brains of diurnal squirrels, and suggest that many of the primate-like features of the visual system of tree shrews arose independently of those in primates.

Introduction

One of the most challenging pursuits in neuroscience is that of reconstructing the evolutionary history of the brains of extant species such as ourselves. The obvious problem in doing so is that brains do not fossilize, and thus little can be learned directly from the fossil record. Although the brain case may be preserved, and the size and the shape of the brain can be closely approximated from the inside dimensions of the brain case, such fossils

KARGER

Fax +41 61 306 12 34
E-Mail karger@karger.ch
www.karger.com

0006–8977/02/0596–0262$18.50/0

Accessible online at:
www.karger.com/journals/bbe

Jon H. Kaas
Department of Psychology, Vanderbilt University
301 David K. Wilson Hall, 111 21st Avenue South
Nashville, TN 37203 (USA)
Tel. +1 615 322 6029, Fax +1 615 322 7941, E-Mail Jon.Kaas@vanderbilt.edu

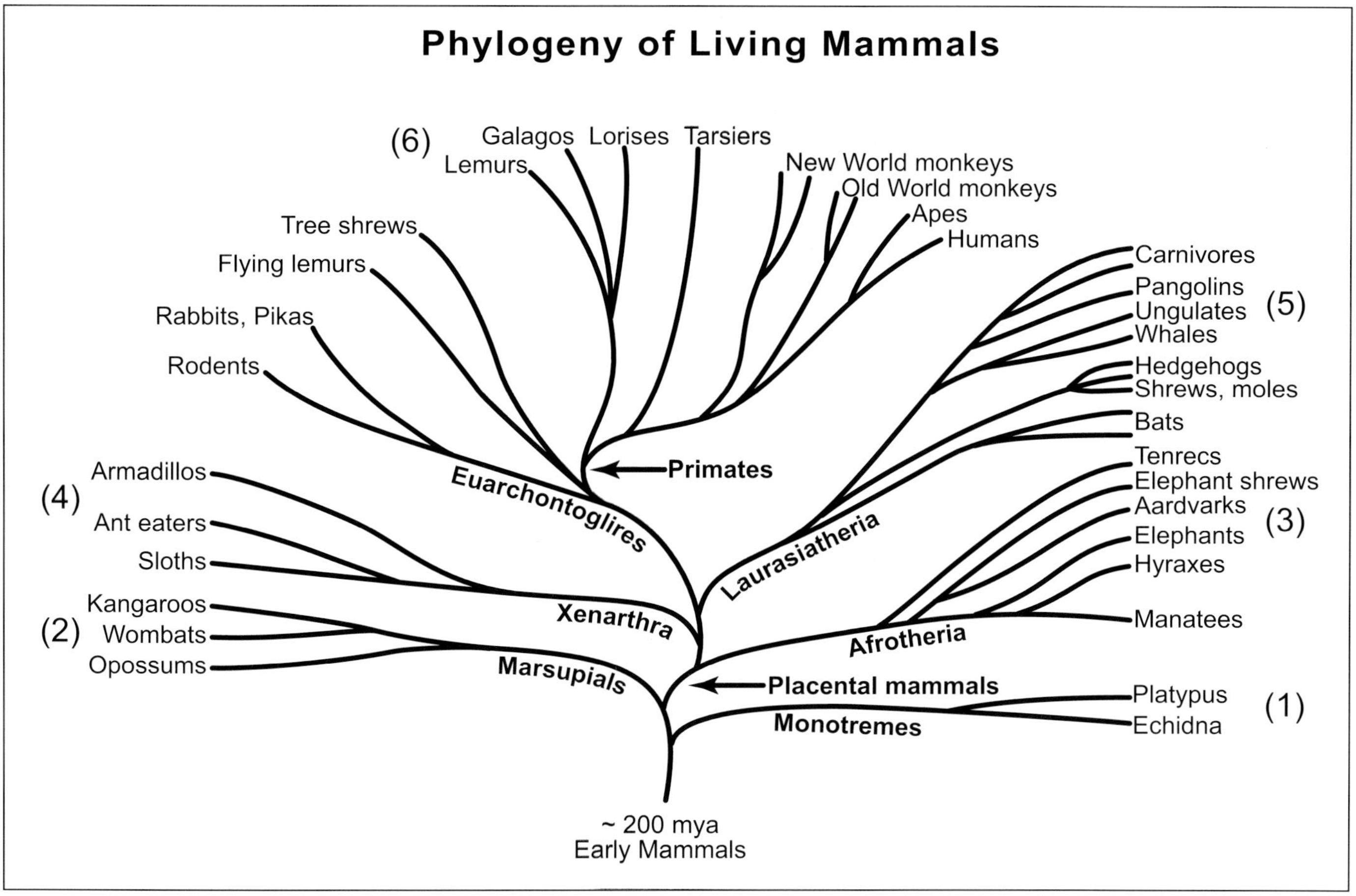

Fig. 1. The phyletic radiation of mammals. Recent attempts to analyze molecular data sets [e.g., Murphy et al., 2001] have led to detailed proposals of how extant mammals are related. Although major disagreements remain, much evidence now supports the division of extant mammals into six major clades or superorders. (1) Monotremes (Prototheria) and (2) marsupials (Metatheria) have long been recognized as early branches of the radiation that are usually distinguished as orders, although marsupials likely contain four or more orders. The oldest placental (Eutherian) superorder, (3) Afrotheria originated in an African part of the fracturing supercontinent. A subsequent superorder (4) Xenarthra, focused in emerging South America. (5) Laurasiatheria includes most insectivores and bats, as well as dogs, cats, whales, cows, and horses. (6) Euarchontoglires contain humans and all our primate relatives, and tree shrews (Scandentia) and flying lemurs (Dermoptera) as close relatives (sister groups). Humans are more distantly related to rabbits and pikas (lagomorphs) and their sister group of rodents (mice, rats, squirrels, beavers).

tell us little about the internal organizations of brains from the past. Thus, it is necessary to infer the organizations of the brains of ancestors from the brains of living species [e.g., Kaas, 1987, 1995; Northcutt and Kaas, 1995; Preuss et al., 1999], where detailed studies of brain organization are possible.

Early efforts to make inferences from living species about brain evolution used a quasi-phylogenetic approach in which related extant species were assigned levels of complexity, and representative species over a series of levels were used as models of brains in an evolutionary sequence. Thus, Eliot Smith [1909] and later, Le Gros Clark [1959] used a sequence of primates at progressively 'higher levels' of brain differentiation to represent the evolutionary history of the human brain [see Preuss, 2000 for further discussion]. For convenience, we call this the 'Clark' approach.

A more modern process is the cladistic approach. This widely-discussed approach especially benefits from the extensive information we now have on the phylogenetic relationships of extant species of mammals [fig. 1; see deJong, 1998; O'Brien et al., 2001; Murphy et al., 2001], and relies on an evaluation procedure known as character analysis [Wagner, 2001]. Species are ordered in a branch-

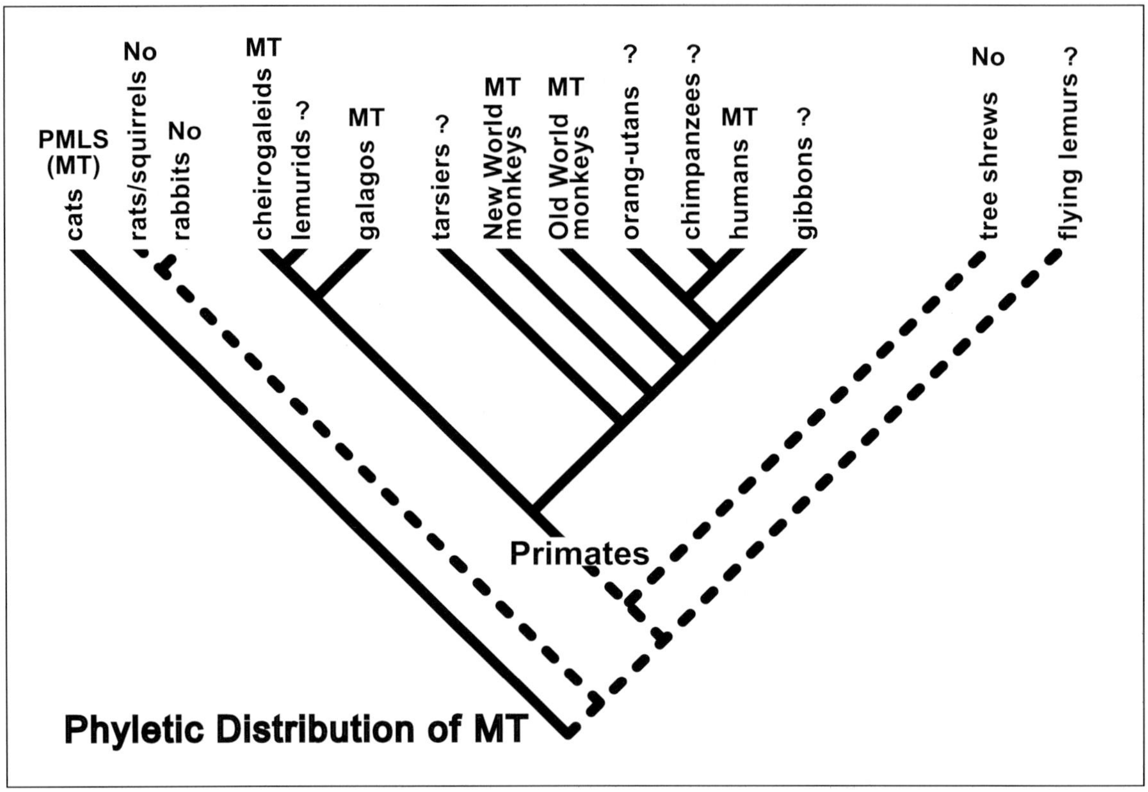

Fig. 2. The phyletic distribution of MT. All studied primates have the middle temporal visual area, MT, although evidence is not yet available for some primates. A visual area in domestic cats, PMLS, is similar to MT in relative location, retinotopic organization, and connections with other visual areas. Thus, PMLS has been considered a homolog of MT. However, a cladistic analysis suggests otherwise, as no MT-like visual area has been found in rodents, lagomorphs or tree shrews.

ing tree (cladogram) according to phylogenetic distance, and the presence or absence of a character (any feature or trait) is determined for as many species as possible across the distribution. Any character that is found in all or most members of a related (monophyletic) group or clade of species is considered to be inherited from a common ancestor and homologous, whereas traits inconsistently distributed within and across clades may be more parsimoniously considered as independently or convergently evolved and homoplasous [e.g., Eldredge and Cracraft, 1980; Wiley, 1981]. This review considers some of the strengths and weaknesses of the 'Clark' and 'Cladistic' approaches, and suggests that an informed Clark approach could have a role in studies of brain evolution.

Examples of the Use of the Cladistic Approach

The usefulness of the cladistic approach in distinguishing homoplasies from homologies can be nicely illustrated with a few examples from the mammalian visual system. Because of the introduction of a number of effective anatomical and physiological techniques, a number of functionally significant subdivisions of visual cortex especially, but also other regions of cortex, have been distinguished within the last 30 years of investigation. One of the first of the newly discovered visual areas was the middle temporal visual area, MT, of monkeys [Allman and Kaas, 1971]. MT is a visual area in the middle of the upper part of the temporal lobe that contains a systematic (retinotopic or visuotopic) representation of the contralateral visual hemifield, receives direct activating inputs from primary visual cortex (V1), has neurons that are highly sensitive to the direction of moving stimuli, and has several architectonic features normally found only in primary sensory areas (high in cytochrome oxidase and myelin). The PMLS area (named by its posteromedial location in the lateral suprasylvian sulcus) in cats has many of these characteristics. Thus, PMLS of cats has been homologized with MT of primates [e.g., Creutzfeldt,

1985; Payne, 1993]. The argument for homology was that there are so many similarities between MT and PMLS that it is unlikely that they emerged independently, and they must reflect an inheritance from a common ancestor. However, a cladistic analysis suggests otherwise. If one constructs a cladogram that includes cats and monkeys (fig. 2), and notes which groups in this cladogram have a MT-like area, we see that all studied primates have an obvious MT, but the closest examined relative of primates, tree shrews, do not. In addition, we see that there is no evidence of a MT-like area in members of the sister taxon (rodents and lagomorphs) to archontan mammals (primates, tree shrews and flying lemurs). Thus, if we have interpreted the neurological evidence correctly, the parsimonious conclusion is that MT and PMLS evolved independently as convergences in the organization of the mammalian visual system. MT apparently emerged with the first primates, but the phyletic distribution of PMLS is still uncertain. Other carnivores need to be examined. No PMLS-like area has been described in shrews, moles, and hedgehogs, but ungulates, pangolins, and whales have not been examined. As all these mammals belong to the Laurasiatherian clade with carnivores (fig. 1), it seems that PMLS may have evolved in only the carnivore radiation. However, a MT-like or PMLS-like area has been claimed for megachiropteran bats [see Rosa, 1999], which also belong to the Laurasiatherian clade.

A cladistic approach also eliminates as homologies a number of other shared features of visual cortex in monkeys and cats. As two examples, primary visual (V1) cortex of monkeys and most other primates is characterized by a distribution of cytochrome oxidase patches (blobs) and by alternating ocular dominance bands (columns), and V1 of cats has both of these features, but they are not found in tree shrews, rodents and lagomorphs [see Preuss and Kaas, 1996]. Thus, they appear to have evolved independently. Although other examples of such convergences could be noted [e.g., see Wicht and Northcutt, 1992; Nishikawa, 1997], the main point is that the cladistic approach has been used with considerable success to distinguish homologies and homoplasies, and construct more probable courses of brain evolution [e.g., Northcutt, 1995].

Difficulties of Using a Cladistic Approach

For the comparative neuroscientist focused on understanding brain evolution, a cladistic approach using character analysis poses some difficulties. Perhaps the most serious problem is that characters are often hard to identify and validate. Some characters, such as the presence or absence of a corpus callosum, are easy to identify because they are obvious to any examiner. The cytochrome oxidase blobs in V1 are another example of an obvious character, although the blobs are not apparent without a suitable processing of brain sections. Thus, the apparent absence of blobs in some primates where the brains have been obtained after death [see McGuinness et al., 1986], might only reflect the difficulty in properly processing such tissue [see Preuss and Kaas, 1996]. The presence or absence of segregated inputs from each eye in primary visual cortex, the ocular dominance columns, seemed easy to detect with a number of experimental procedures, but recently new evidence for such columns has been obtained in primates thought to lack them [Horton and Hocking, 1996; Chappert-Piquemal et al., 2001]. A further complication is that ocular dominance columns are weakly or strongly expressed in different primates. The character state is not simply present or absent, but intermediate conditions exist. Thus, it seems reasonable to reconsider previous conclusions that such columns evolved more than once in primates [e.g., Florence et al., 1986] in favor of the possibility that a weak expression of columns (or a sensitive substrate for columns) in early primates has evolved into a strong expression more than once.

The presence or absence of a third visual area in extrastriate cortex, V3, is also difficult to address in a cladistic manner. Although the evidence for the existence of a V3 has long been accepted for cats [see Albus and Beckman, 1980], this has not been the case for primates. Currently, the organization of cortex in the V3 region of monkeys is being interpreted differently as possibly containing V3, or having a V3 that is divided into two separate visual areas (dorsal V3 and the ventral posterior area, VP), or having other visual areas in the location of V3 [see Kaas and Lyon, 2001]. More compelling evidence for V3 in monkeys has been obtained only recently [Lyon and Kaas, 2001; 2002], even though there have been many years of extensive study. Thus, it is now possible to construct a cladogram of the distribution of V3 that supports the conclusion that V3 evolved independently in cats and primates (fig. 3). However, as V3 has been so difficult to validate in monkeys and prosimians, one wonders about the assumption that V3 does not exist in tree shrews and other mammals where evidence for V3 is lacking. Contrary to other proposals [Kaas et al., 1989], a V3 has been claimed for squirrels [Sereno et al., 1991], but the evidence has not been published. Yet, compelling evidence could be published at any time. In brief, the cladistic distribution of a V3, or any other cortical area, can be extremely difficult to determine. Cortical areas are usually established as valid

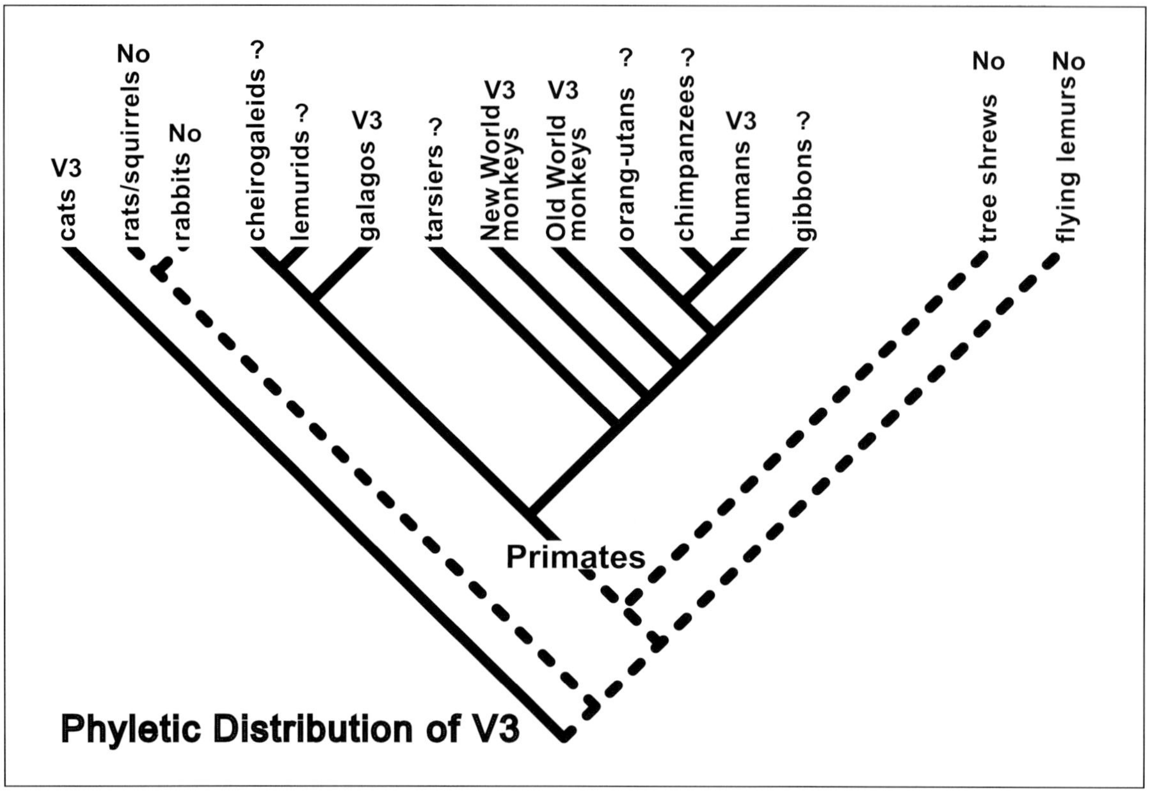

Fig. 3. The phyletic distribution of the third visual area, V3. A visual area along the outer border of the second visual area, V2, in cats has been termed V3 (VIII). A highly similar area has been described in monkeys, but other interpretations have remained viable. Recently, more evidence has accumulated and a 'V3' has been identified in galagos, several species of New World monkeys, and Old World macaque monkeys. Non-invasive imaging data from humans are highly consistent with the presence of V3. Other primates have not been adequately examined. However, studies of rabbits, rats, squirrels, and tree shrews have provided little evidence for V3 in these groups, its suggesting that a V3-like area has evolved independently in primates and carnivores.

by compatible evidence of different types, and new interpretations can emerge with new evidence.

A related problem is that many characters may be identified only with experimental procedures, and many species are protected and unavailable for experimental study. If the detection of V3, for example, depends on revealing patterns of V3 connections and retinotopic organization, we have no present way of obtaining this information from many prosimian primates, tarsiers, and great apes, and only non-invasive imaging of retinotopy can be obtained from humans [where the retinotopy is consistent with the concept of V3, but open to other interpretations; see Kaas and Lyon, 2001].

Finally, comparative studies can be costly and labor intensive, so that even if they are feasible, only a few species are likely to be examined. Indeed, neuroscience research has concentrated on a few laboratory species, with most current efforts on rats and mice. Because of the funding structure, we are largely ignorant about those features of brain organization that require experimental verification for most mammals. In brief, the cladistic approach is obviously powerful, but its application to the reconstruction of the courses of brain evolution has been very limited because of the above constraints. When cladograms contain few species, they may not look much different from Le Gros Clark's series of representative mammals at different levels.

Problems of the Clark Approach

The weaknesses of the Clark approach are obvious, and they have been outlined elsewhere [Hodos and Campbell, 1969; Preuss, 1995; Striedter, 1998]. The brains of extant mammals are various mosaics of primitive and derived features. More importantly, new features, such as cortical areas and modules may emerge, but their functional integration into a complex neural system means that many other parts of the brain are thereby modified. The functionally significant subdivisions of the brain (nuclei, areas, modules, circuits), are unlikely to be completely primitive or completely derived. Thus, the 'triune brain' theory [MacLean, 1990], proposing that we have retained major components of our brains from reptilian and other ancestors, is a simplification that might be instructive, but it does not reflect the reality of extensive modification [see Butler and Hodos, 1996, p 86]. A major problem with the Clark approach is that a single direction of change for all structures is assumed without compelling evidence. Undifferentiated, simple, and small brain structures are not necessarily primitive, and brain characters can be lost as well as gained [see Catania et al., 1999]. Although some theories of brain evolution are based on the assumption that new and old brain features can be identified in human and other brains by a less differentiated architectonic appearance [e.g., Sanides, 1970], there is little justification for such an assumption [see Kaas, 1987; Butler and Hodos, 1996]. Another major weakness of the Clark approach is that convergent specializations are not easily detected, and similarities could be wrongly attributed to inheritance from a common ancestor.

A Truncated Cladistic Approach

Although the value of a cladistic approach is obvious, practical concerns mean that theories of brain evolution for any line of descent will be based largely on observations from a few species. We need to compare, but for practical reasons we may only be able to study a few species and we need ways of choosing. Possibly species will be chosen by some of the same criteria used by Le Gros Clark. It seems reasonable, for example, to suppose that the brains of extant prosimians have changed less overall from those of early primates than those of apes, if only because they are similar in overall size and shape [Ross, 2000]. Given this premise, brain traits found in galagos but not chimpanzees are more likely to resemble those of the common ancestor than brain traits found in chimpanzees but not galagos. In reality, inferences about the evolution of the human brain depend greatly on evidence from experimental studies on macaque monkeys, several species of New World monkeys, prosimian galagos, and tree shrews, the closest relative of primates available for experimental study. This series of representative species with brains at successive 'levels' of complexity, extended by more limited information from chimpanzees and humans at one end and from insectivores such as hedgehogs and tenrecs (representing early Eutherian mammals) at the other, is very much like the Le Gros Clark series, and it suffers from the same weaknesses. However, we can compensate for the weaknesses in several ways. First, we can extend our range of comparisons wherever possible, and these comparisons can be guided by current theories of mammalian phylogeny [fig. 1; see de Jong, 1998; Murphy et al., 2001; O'Brien et al., 2001]. For example, now that we know that the insectivore hedgehogs and tenrecs are not closely related, we can more properly assume that their shared brain features reflect those retained from early placental mammals. Second, even limited information on some species can be quite valuable. Often, brain features that have been established in one species by extensive experimental evidence can be detected and established by much more limited evidence in another species. For example, once MT was established as a visual area in monkeys by a characteristic pattern of retinotopic organization, connections with other visual areas, and architectonic features, it became possible to demonstrate the probable existence of MT in protected primates by the presence of a diagnostic histochemical characteristic in brains examined after death [e.g., Preuss and Kaas, 1996]. Perhaps histochemical markers for other visual areas such as V3 will be discovered. Third, we can evaluate the probability of convergences in brain features by considering how the brains of distantly related species with similar niches resemble each other. To the extent that brain features appear to be adaptations to specific environments, we need to be cautious about attributing them to ancestral states, especially when the ancestor occupied a different niche. Below, we see how a search for convergent brain structures as adaptations to similar environments can help guide theories of brain evolution.

Comparing Tree Shrews and Squirrels

Natural selection shapes life forms to particular ecological niches. As a result, remarkably similar species of flightless birds, ant eaters, burrowing insectivores, and grazers have evolved independently on the African, South American, and Australian land masses. The hummingbird moth looks and acts very much like a hummingbird, and legless lizards appear to be snakes. The apparent predictability of the general characteristics of how mammals and birds adapt to specific niches led Dougal Dixon [1981] to imaginatively speculate on the forms that would evolve in the 50 million years that would follow the next

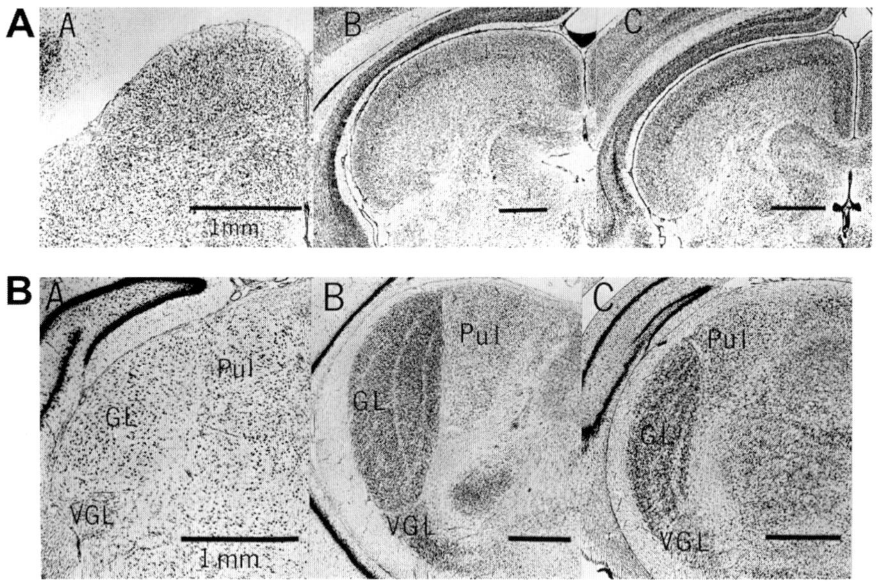

Fig. 4. Comparison of the subcortical visual system of (A) hedgehogs, (B) squirrels and (C) tree shrews. **A** Frontal brain sections through the superior colliculus (SC) of the midbrain (Nissl stain). Note the large size and distinct lamination of the SC in grey squirrels (B) and tree shrews (C) compared to the hedgehog (A). **B** Frontal brain sections through the dorsal lateral geniculate nucleus (GL) of the thalamus. The GL is distinctly, but differently, laminated in squirrels (B) and tree shrews (C), whereas hedgehogs (A) have a poorly differentiated GL with no visible lamination. Note the similarity also in the visual pulvinar (Pul) in tree shrews and squirrels, as well as in the ventral lateral geniculate nucleus (VGL).

massive extinction of most life, as we ruin the environment. There is no way we can evaluate this daring step into predictive biology, but we can evaluate degrees of similarities between distantly related extant species that have independently adapted to similar niches. Do such species have similar brains, and if so, can we make informed statements about the brains of extant species if we know something about their environmental niches? Can we use such information to identify specializations and indicate direction of change, and thereby guide theories based on model species at 'successive' levels? As the closest relative of primates that is available for experimental study (fig. 1), features of tree shrew brains are often compared to those of primate brains. However, how can brain modifications related to the specific niche of tree shrews be distinguished from those shared by an ancestor of tree shrews and primates, thereby guiding and constraining the use of tree shrews as the Clark model of a protoprimate?

One approach to this problem is to compare the brains of tree shrews to those of mammals occupying a similar niche. The first tree shrew collected by a Western naturalist in 1780 was thought to be a squirrel, and forty years passed before tree shrews were recognized as different from squirrels [Emmons, 2000]. In the Malay/Indonesian language, tree shrews and squirrels are known by the same name, tupai. They are both agile, diurnal, semiarboreal mammals with large eyes, about 60° of binocular vision, and acute visual awareness. Among their many similarities, tree shrews and squirrels both use a tail flick to signal danger [Emmons, 2000]. As tree shrews and squirrels are convergent in body form, emphasis on vision, and other aspects of behavior, it seems that they would likely have convergent brain features as well.

As expected, the visual systems of tree shrews and squirrels display a number of similarities. In both, the superior colliculus, the major visual center of the midbrain, is unusually large and distinctly laminated [Abplanalp, 1970; Lane et al., 1971] compared to the superior colliculus of most mammals, including all primates [Kaas and Huerta, 1988]. The specialization of this structure is especially obvious when compared to the small, poorly differentiated superior colliculus of a nocturnal hedgehog (fig. 4), which displays a colliculus that is similar in appearance to that of a wide range of mammals with small eyes and little emphasis on vision. In a rat and a ground squirrel of similar body size, the surface area of the superior colliculus in the squirrel is about 10 times greater than the rat [Kaas and Collins, 2001]. Thus, rodents vary considerably in the appearance of the superior colliculus, and squirrels resemble tree shrews much more closely in this feature than they resemble other rodents such as rats and mice. It seems reasonable to conclude from such comparative observations that the large size and distinct lamination of the superior colliculus in tree shrews are not retained from a common ancestor with primates, but are adaptations of tree shrews and squirrels. Early primates also emphasized vision, but in contrast to tree shrews and squirrels, they appear to have been nocturnal [Ross, 2000].

If we consider the other major target of the retinal ganglion cells, the dorsal lateral geniculate nucleus (LGN) of the visual thalamus, tree shrews and squirrels again are convergent. Compared to the LGN of a hedgehog, the LGN's of squirrels and tree shrews are large, distinct from other parts of the thalamus, and conspicuously laminated (fig. 4). However, tree shrews have more and narrower layers; the layers in squirrels and tree shrews do not segregate functional classes of neurons in the same way [Kaas et al., 1972a; Norton, 1982; Conway and Schiller, 1983]. The adjacent pulvinar nucleus (or nuclei) of the visual thalamus is also large and very distinct in both tree shrews and squirrels. Although the pulvinar appears to be superficially similar in both mammals, different subdivisions may exist [Lyon et al., 1997]. The differentiation and enlargement of the visual thalamus appears to be an adaptation for both tree shrews and squirrels, although the details of the differentiation are not the same.

At the level of visual cortex, additional convergences are apparent. Most notably, primary visual cortex (V1) is large (fig. 5), and histologically distinct (fig. 4). However, the sublayers of layer 4 are physiologically and anatomically different [Kretz et al., 1986; Conley et al., 1994]. Much more of V1 is devoted to the central few degrees of vision just in front of the nose [Hall et al., 1971; Kaas et al., 1972b, c] than in most mammals, but not as much as in primates. The second area, V2, is also architectonically distinct, and a number of other visual areas occupy occipital and temporal cortex lateral to V2 (fig. 5). In both mammals, the temporal cortex is greatly expanded to accommodate these visual areas. Again, these similar features differ from those in mammals with more generalized visual systems, where visual areas occupy proportionately less of the cortex, visual areas are less distinct histologically, and there are fewer visual areas.

As highly visual mammals, tree shrews and squirrels devote proportionately less cortex to auditory and somatosensory inputs, and there is little frontal cortex in addition to that devoted to primary motor and premotor areas. Both mammals manipulate food and other objects with the forepaws, and a large forepaw representation is apparent in somatosensory cortex [Sur et al., 1978; 1980; Krubitzer et al., 1986].

Also, there are notable differences in cortical organization between squirrels and tree shrews. The temporal lobes are differently divided into areas [fig. 5; Kaas et al., 1972b, 1989; Sesma et al., 1984; Lyon et al., 1998]. Such differences indicate that the adaptations to similar niches are expressed broadly, and the details of the patterns differ, reflecting the forebrains of recent, less similar ancestors and possibly the moderate but perhaps significant differences in niche.

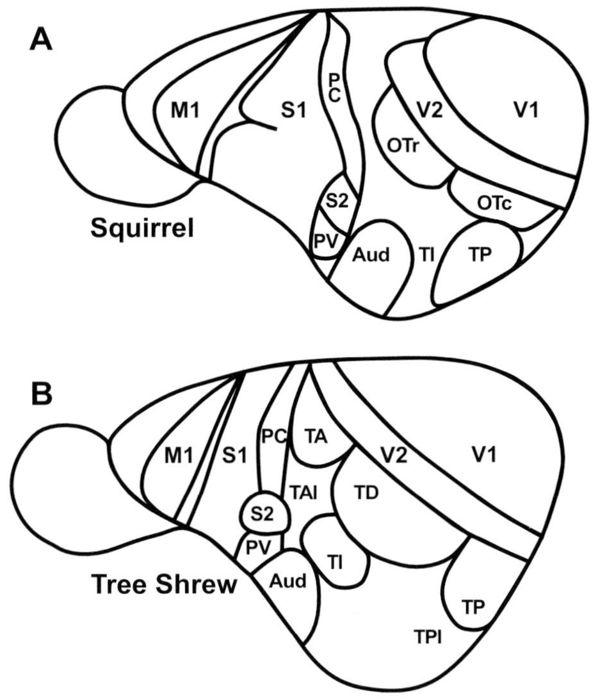

Fig. 5. Dorsolateral views of the brains of a (**A**) squirrel and a (**B**) tree shrew with proposed subdivisions of cortex. Note that both have an expanded temporal region of the cerebral hemisphere, reflecting a shared emphasis on vision. In both, primary (V1) and secondary (V2) visual areas are large, and the temporal lobes contain a number of additional visual areas. However, temporal visual areas appear to be differently subdivided into visual areas. Visual areas include occipital temporal (OT) areas that are rostral (OTr) or caudal (OTc), and temporal intermediate (TI) and temporal posterior (TP) for squirrels [see Kaas et al., 1989]. Proposed visual areas for tree shrews [see Lyon et al., 1998] include temporal anterior (TA), dorsal (TD), and posterior (TP) areas as well as temporal inferior (TI) and temporal anterior inferior (TAI) and temporal posterior inferior (TPI) areas. A primary motor area (M1), somatosensory areas (S1, S2, PV, PC), and an auditory (Aud) region [see Luethke et al., 1988] are indicated for both tree shrews and squirrels.

What Does This Imply about the Forebrains of Early Primates?

There are a number of features of forebrain organization that tree shrews, squirrels, and many primates share. These are likely to have emerged because all three groups occupy semi-arboreal or arboreal niches where frontal vision is important, and objects are manipulated with the forepaws. In all, the LGN is laminated, the superior colliculus is enlarged, primary visual cortex is distinctly layered, and a number of extrastriate visual areas exist in an expanded temporal lobe. Somatosensory cortex is only moderately differentiated, but the glabrous skin of the forepaw is well-represented. Such similarities reinforce the concept that early primates, as small, visual climbers also had many of these shared features as adaptations. However, some shared features of tree shrews and squirrels would likely not have been present in early primates. The superior colliculus is proportionately larger and more differentiated in tree shrews and squirrels than in extant primates, and this suggests that early primates had a smaller superior colliculus. These shared differences from primates in the visual system may reflect the important distinction that early primates were nocturnal [Ross, 2000], whereas tree shrews [Emmons, 2000] and most tree squirrels [Walls, 1942] are highly diurnal (although flying squirrels have become nocturnal). Thus, the superior colliculus would have been smaller and less differentiated in early nocturnal primates as in other nocturnal mammals. In addition, both tree shrews and squirrels retain the generalized feature of representing the complete contralateral eye in the superior colliculus, whereas all extant primates represent only the contralateral hemifield in the superior colliculus [see Kaas and Preuss, 1993]. The LGN of early primates would likely be laminated, but the arrangements of layers and the segregations of functional types of neurons in layers would be different than in tree shrews, as different lamination patterns are seen in all three groups. All extant primates have separate layers for the 'M' and 'P' classes of retinal ganglion cell inputs, and thus early primates probably also had this feature. Visual areas retained from the first mammals, notably V1 and V2, are expanded and differentiated in tree shrews, squirrels, and primates, and all have a number of extrastriate visual areas. However, different additions likely characterize each group, and it is uncertain if tree shrews and primates share any visual areas in addition to V1 and V2.

We can conclude, as their similar body forms, behaviors, and diurnal niches suggest, that the organizations of the forebrains of tree shrews and squirrels have become highly similar overall, while remaining different in detail. Although tree shrews (Scandentia) are much more closely related to primates than squirrels (Rodentia), the visual system of tree shrews more closely resembles that of squirrels. Remarkably, there is no clear evidence of any feature of the tree shrew visual system that is shared with primates, but not squirrels, although there may be many. Le Gros Clark [1959] saw the well-differentiated and enlarged visual system of tree shrews as evidence that tree shrews should be classified as primates, as both groups seemed to have inherited a well-developed visual system from a common (early primate) ancestor. Although this still may be the case (but from a non-primate ancestor), we see no evidence of features retained from such an ancestor, and only evidence that selection for the small-mammal, diurnal, semiarboreal niche was a powerful force in shaping the components of the visual system of tree shrews [also, see Campbell, 1980]. The selection pressures on early emerging primates were somewhat similar, in that they were also small, semiarboreal or arboreal, visually dependent mammals, but they were also nocturnal visual predators [Ross, 2000]. Common features of selection resulted in enlarged, well-differentiated visual systems in primates, tree shrews, and squirrels, but the differences across groups are obvious with closer inspection. In particular, the emphasis on the superior colliculus as a prominent visual system was less pronounced in primate evolution, possibly reflecting the early nocturnal niche. Surprisingly, the later emerging, diurnal, anthropoid primates of small size (e.g., marmosets) did not evolve a tree shrew-like visual midbrain. Tree shrews evolved segregated processing streams for the 'on' and 'off' ganglion cell classes of the retina, but primates evolved segregated streams for 'M' and 'P' cells. It is intriguing to consider these different specializations as related to the differing requirements of diurnal and nocturnal niches, with a commitment to the early nocturnal pattern in subsequently diurnal primates.

In the end, we are forced to conclude that the brain of a squirrel would serve almost as well as a model of the early protoprimate brain, not because of a close phylogenetic relationship, but simply because of convergences in brain organization driven by selection for somewhat similar niches. The many convergences in the squirrel and tree shrew visual systems demonstrate the danger in assuming that similarities in tree shrews and primate visual systems are retentions from a common ancestor, rather than homoplasies, and this is an important lesson. In addition, this comparison suggests that we might deduce much about primate brain evolution from studies of distantly related mammals adapted to primate-like niches. Yet this does not mean that it will be unproductive to study brain organization in tree shrews to further understanding of primate brain evolution. However, we will need to be very careful about distinguishing convergences from homologies. Unfortunately, cladistic efforts to do so will be constrained by the evidence that primates are only closely related to tree shrews, which are available for study, and flying lemurs (Dermoptera), which are not. Additional comparisons will depend on much more distantly related mammals (fig. 1).

Acknowledgments

Helpful comments on an earlier draft of this paper were given by Leah Krubitzer and Todd Preuss, and by the volume editors, Heather Eisthen and Kiisa Nishikawa.

References

Abplanalp, P. (1970) Some subcortical connections of the visual system in tree shrews and squirrels. Brain Behav. Evol., *3:* 155–168.

Albus, K., and R. Beckman (1980) Second and third visual areas of the cat: Interindividual variability in retinotopic arrangement and cortical location. J. Physiol. (London), *299:* 247–276.

Allman, J.M., and J.H. Kaas (1971) Representation of the visual field in striate and adjoining cortex of the owl monkey *(Aotus trivigatus)*. Brain Res., *35:* 89–106.

Butler, A.B., and W. Hodos (1996) Comparative Vertebrate Neuroanatomy. Wiley-Liss, New York.

Campbell, C.B.G. (1980) The nervous system of the tupaiidae: Its bearing on phyletic relationships. *In* Comparative Biology and Evolutionary Relationships of Tree Shrews (ed. by W.D. Luckett), Plenum Press, New York, pp. 219–242.

Catania, K.C., D.C. Lyon, O.B. Mock, and J.H. Kaas (1999) Cortical organization in shrews: Evidence from five species. J. Comp. Neurol., *410:* 55–72.

Chappert-Piquemal, C., C. Fonta, and F. Malecaze (2001) Ocular dominance columns in the adult New World monkey, *Callithrix jacchus*. Vis. Neurosci., *18:* 407–412.

Clark, Le Gros W.E. (1959) The Antecedents of Man. Edinburgh University Press, Edinburgh, UK.

Conley, M., D. Fitzpatrick, and I.T. Diamond (1984) The laminar organization of the lateral geniculate body and the striate cortex in the tree shrew *(Tupaia glis)*. J. Neurosci., *4:* 171–197.

Conway, J.L., and P.H. Schiller (1983) Laminar organization of tree shrew dorsal lateral geniculate nucleus. J. Neurophysiol., *50:* 1330–1342.

Creutzfeldt, O. (1985) Comparative aspects of representation in the visual system. Exp. Brain Res., Suppl., *11:* 53–81.

deJong, W.W. (1998) Molecules remodel the mammalian tree. Trends Ecol. Evol., *13:* 270–275.

Dixon, D. (1981) After Man: A Zoology of the Future. St. Martin's Press, New York.

Eldredge, N., and J. Cracraft (1980) Phylogenetic Patterns and the Evolutionary Process. Columbia University Press, New York.

Emmons, L.H. (2000) Tupai: A Field Study of Bornean Tree Shrews. University of California Press, Berkeley., CA.

Florence, S.L., M. Conley, and V.A. Casagrande (1986) Ocular dominance columns and retinal projections in New World spider monkeys *(Ateles ater)*. J. Comp. Neurol., *243:* 234–248.

Hall, W.C., J.H. Kaas, H. Killackey, and I.T. Diamond (1971) Cortical visual areas in grey squirrel *(Scuirus carolinensis)*: A correlation between evoked potential maps and architectonic subdivisions. J. Neurophysiol., *34:* 437–452.

Hodos, W., and C.B.G. Campbell (1969) *Scala naturae*: Why there is no theory in comparative psychology. Psychol. Rev., *76:* 337–350.

Horton, J.C., and D.R. Hocking (1996) Anatomical demonstration of ocular dominance columns in striate cortex of the squirrel monkey. J. Neurosci., *16:* 5510–5520.

Kaas, J.H. (1987) The organization and evolution of neocortex. *In* Higher Brain Functions (ed. by S.P. Wise), John Wiley and Sons, New York, pp. 347–378.

Kaas, J.H. (1995) The evolution of isocortex. Brain Behav. Evol., *46:* 187–196.

Kaas, J.H., and C.E. Collins (2001) Variability in sizes of brain parts. Behav. Brain Sci., *24:* 288–290.

Kaas, J.H., and M.F. Huerta (1988) Subcortical visual system of primates. *In* Comparative Primate Biology, Volume 4: Neurosciences (ed. by H.P. Steklis), Alan R. Liss, New York, pp. 327–391.

Kaas, J.H., and D.C. Lyon (2001) Visual cortex organization in primates: Theories of V3 and adjoining visual areas. Prog. Brain Res., *134:* 285–295.

Kaas, J.H., and T.M. Preuss (1993) Archontan affinities as reflected in the visual system. *In* Mammal Phylogeny (ed. by F. Szalay, M. Novacek and M. McKenna), Springer-Verlag, New York, pp. 115–128.

Kaas, J.H., R.W. Guillery, and J.M. Allman (1972a) Some principles of organization in the dorsal lateral geniculate nucleus. Brain Behav. Evol., *6:* 253–299.

Kaas, J.H., W.C. Hall, and I.T. Diamond (1972b) Visual cortex of the grey squirrel: Architectonic subdivisions and connections from the visual thalamus. J. Comp. Neurol., *145:* 273–306.

Kaas, J.H., W.C. Hall, H. Killackey, and I.T. Diamond (1972c) Visual cortex of the tree shrew *(Tupaia glis)*: Architectonic subdivisions and representations of the visual field. Brain Res., *42:* 491–496.

Kaas, J.H., L.A. Krubitzer, and K.L. Johanson (1989) Cortical connections of area 17 (V1) and 18 (V-II) of squirrels. J. Comp. Neurol., *281:* 426–446.

Kretz, R., G. Rager, and T.T. Norton (1986) Laminar organization of ON and OFF regions and ocular dominance in the striate cortex of the tree shrew *(Tupaia belangeri)*. J. Comp. Neurol., *251:* 135–145.

Krubitzer, L.A., M.A. Sesma, and J.H. Kaas (1986) Microelectrode maps, myeloarchitecture, and cortical connections of three somatotopically organized representations of the body surface in parietal cortex of squirrels. J. Comp. Neurol., *250:* 403–430.

Lane, R.H., J.M. Allman, and J.H. Kaas (1971) Representation of the visual field in the superior colliculus of the grey squirrel *(Sciurus carolinesis)* and the tree shrew *(Tupaia glis)*. Brain Res., *26:* 277–292.

Luethke, L.E., L.A. Krubitzer, and J.H. Kaas (1988) Cortical connections of electrophysiologically and architectonically defined subdivisions of auditory cortex in squirrels. J. Comp. Neurol., *268:* 181–203.

Lyon, D.C., and J.H. Kaas (2001) Connectional and architectonic evidence for dorsal and ventral V3 and area DM in marmoset monkeys. J. Neurosci., *21:* 249–261.

Lyon, D.C., and J.H. Kaas (2002) Evidence for a modified V3 with dorsal and ventral halves in macaque monkeys. Neuron, *33:* 453–461.

Lyon, D.C., N. Jain, and J.H. Kaas (1997) Callosal and thalamic connections of the extrastriate visual cortex in tree shrews. Soc. Neurosci. Abstr., *23:* 1031.

Lyon, D., N. Jain, and J.H. Kaas (1998) Cortical connections of striate and extrastriate visual areas in the tree shrew. J. Comp. Neurol., *401:* 109–128.

MacLean, D.D. (1990) The Triune Brain in Evolution. Plenum Press, New York.
McGuinness, E., C. McDonald, M. Sereno, and J.M. Allman (1986) Primates without blobs: The distribution of cytochrome oxidase activity in *Tarsius*, *Hapalemur*, and *Cheirogaleus*. Soc. Neurosci. Abstr., *12:* 130.
Murphy, W.J., E. Eizirik, S.J. O'Brien, O. Madsen, M. Scally, C.J. Douady, E. Teeling, O.A. Ryder, M.J. Stanhope, W.W. DeJong, and M.S. Springer (2001) Resolution of the early placental mammal radiation using Bayesian phylogenetics. Science, *294:* 2348–2351.
Nishikawa, K.C. (1997) Emergence of novel functions during brain evolution. BioScience, *47:* 341–354.
Northcutt, R.G. (1995) The forebrain of gnathostomes: In search of a morphotype. Brain Behav. Evol., *46:* 275–318.
Northcutt, R.G., and J.H. Kaas (1995) The emergence and evolution of mammalian neocortex. Trends Neurosci., *18:* 373–379.
Norton, T.T. (1982) Geniculate and extrageniculate visual systems in the tree shrew. *In* Changing Concepts of the Nervous System (ed. by A.R. Morrison and P.L. Strick), Academic Press, New York, pp. 377–409.
O'Brien, S.J., E. Eizirik, and W.J. Murphy (2001) On choosing mammalian genomes for sequencing. Science, *292:* 2264–2266.
Payne, B.R. (1993) Evidence for visual cortical area homologs in cat and macaque monkey. Cerebral Cortex, *3:* 1–25.
Preuss, T.M. (1995) The argument from animals to humans in cognitive neuroscience. *In* The Cognitive Neurosciences (ed. by M.S. Gazzaniga), MIT Press, Cambridge, MA, pp. 1227–1241.
Preuss, T.M. (2000) Taking the measure of diversity: Comparative alternatives to the model-animal paradigm in cortical neuroscience. Brain Behav. Evol., *55:* 287–299.
Preuss, T.M., and J.H. Kaas (1996) Cytochrome oxidase 'blobs' and other characteristics of primary visual cortex in a lemuroid primate, *Cheirogaleus medius*. Brain Behav. Evol., *47:* 103–112.
Preuss, T.M., H.S. Qi, and J.H. Kaas (1999) Distinctive compartmental organization of human primary visual cortex. Proc. Nat. Acad. Sci. USA, *96:* 11601–11606.
Rosa, M.G.P. (1999) Topographic organization of extrastriate areas in the flying fox: Implications for the evolution of mammalian visual cortex. J. Comp. Neurol., *411:* 503–523.
Ross, C.F. (2000) Into the light: The origin of Anthropoidea. Ann. Rev. Anthropol., *29:* 147–194.
Sanides, F. (1970) Functional architecture of motor and sensory cortices in primates in the light of a new concept of neocortex evolution. *In* The Primate Brain (ed. by C.R. Noback and W. Montagna), Appleton-Century-Crofts, New York, pp 137–208.
Sereno, J.I., H.R. Rodman, and H.J. Karten (1991) Organization of visual cortex in the California ground squirrel. Soc. Neurosci. Abstr., *17:* 844.
Sesma, M.A., V.A. Casagrande, and J.H. Kaas (1984) Cortical connections of area 17 in tree shrews. J. Comp. Neurol., *230:* 337–351.
Smith, E.G. (1924) The Evolution of Man. Essays. Oxford University Press, London, UK.
Striedter, G.F. (1998) Progress in the study of brain evolution: From speculative theories to testable hypotheses. Anat. Rec. (New Anat.), *253:* 105–112.
Sur, M., R.J. Nelson, and J.H. Kaas (1978) The representation of the body surface in somatosensory area I of the grey squirrel. J. Comp. Neurol., *179:* 425–450.
Sur, M., R.E. Weller, and J.H. Kaas (1980) The body surface representation in somatosensory area I of the tree shrew *(Tupaia glis)*. J. Comp. Neurol., *194:* 71–95.
Wagner, G.P. (2001) The character concept in evolutionary biology. Academic Press, San Diego, CA.
Walls, G.L. (1942) The Vertebrate Eye. Cranbrook Institute of Science, Bloomfield Hills, Michigan, MI.
Wicht, H., and R.G. Northcutt (1992) The forebrain of the Pacific hagfish: A cladistic reconstruction of the ancestral craniate forebrain. Brain Behav. Evol., *40:* 25–64.
Wiley, E.O. (1981) Phylogenetics: The Theory and Practice of Phylogenetic Systematics. John Wiley and Sons, New York.

Brain, Behavior and Evolution

Brain Behav Evol 2002;59:273–293

Why Are Olfactory Systems of Different Animals So Similar?

Heather L. Eisthen

Department of Zoology, Michigan State University, East Lansing, Mich., USA

Key Words

Convergence · Adaptation · Constraint · Odorant binding protein · G Protein-coupled receptor · Sensory transduction · Glomerulus

Abstract

As we learn more about the neurobiology of olfaction, it is becoming increasingly clear that olfactory systems of animals in disparate phyla possess many striking features in common. Why? Do these features provide clues about the ways the nervous system processes olfactory information? This might be the case if these commonalities are convergent adaptations that serve similar functions, but similar features can be present in disparate animals for other reasons. For example, similar features may be present because of inheritance from a common ancestor (homology), may represent responses to similar constraints, or may be superficial or reflect strategies used by researchers studying the system. In this paper, I examine four examples of features of olfactory systems in members of different phyla: the presence of odorant binding proteins in the fluid overlying olfactory receptor neurons; the use of G protein-coupled receptors as odorant receptors; the use of a two-step pathway in the transduction of odorant signals; and the presence of glomerular neuropils in the first central target of the axons of olfactory receptor cells. I analyze data from nematodes, arthropods, molluscs, and vertebrates to investigate the phylogenetic distribution of these features, and to try to explain the presence of these features in disparate animals. Phylogenetic analyses indicate that these features are not homologous across phyla. Although these features are often interpreted as convergent adaptations, I find that alternative explanations are difficult to dismiss. In many cases, it seems that olfactory system features that are similar across phyla may reflect both responses to similar constraints and adaptations for similar tasks.

Introduction

Striking similarities in the organization of vertebrate and insect olfactory systems have long been noted by biologists. As researchers have broadened the scope of inquiry to include nematodes, molluscs, and crustaceans, the impression that some features are consistently present in olfactory systems has only been strengthened. What is the underlying cause of these similarities? Do they represent elements that are essential for olfaction, or are there other explanations for their appearance in distantly-related phyla? In this paper, I will describe features that are

KARGER

Fax +41 61 306 12 34
E-Mail karger@karger.ch
www.karger.com

0006–8977/02/0596–0273$18.50/0

Accessible online at:
www.karger.com/journals/bbe

Heather L. Eisthen
Department of Zoology
Michigan State University, 203 Natural Science Building
East Lansing, MI 48824 (USA)
Tel. +1 517 353 1953, Fax +1 517 432 2789, E-Mail eisthen@msu.edu

present in the olfactory system of members of several phyla, and will analyze possible explanations for the existence of these similarities.

Some researchers who take a 'model system' approach pay little attention to evolutionary issues, and apparently assume that similarities in the development, anatomy, physiology, or behavior of different animals reflect basic principles of biology. Others assume that the features of interest arose once, perhaps in a distant ancestor of the animals being compared; this similarity due to shared inheritance is called *homology*. Similar features might also have evolved independently, through *convergence*. (In this paper, I will consider *parallelism*, in which similar features evolve independently from the same precursor, to be a special case of convergence, as the two are conceptually related and are difficult to distinguish in practice [Eldredge and Cracraft, 1980].) The possibility that similarities in the organization of olfactory systems of diverse animals are due to convergence, and therefore might provide clues about mechanisms of processing of odorant information, has received increased attention in recent years [Ache, 1994; Hildebrand and Shepherd, 1997; Strausfeld and Hildebrand, 1999; Ache and Restrepo, 2000].

Convergent features can arise through different processes, and we should not assume that features that have arisen independently represent adaptations for processing odorant information. In some cases, similar features could arise independently due to *constraints* rather than as a result of adaptation [Wake, 1991]. Constraints can be attributed to many causes, and the problem of constraint has been the subject of many excellent reviews [Maynard Smith et al., 1985; Schwenk, 1994/95]. Developmental and genetic processes constitute an important source of constraint, as these factors can limit the possible phenotypes that can be produced. For example, Finlay and Darlington [1995] have argued that, in general, the cellular events in early neural development constrain size changes such that selection for an increase in size of a particular brain region in mammals will result in an increase in overall brain size, rather than an increase in the size of just that particular region. Constraints are also imposed by physical laws, such as those governing the transmission of sensory stimuli or the biomechanical properties of animals' bodies. Physical laws similarly constrain the architecture of the nervous system, as illustrated by Ringo's [1991] analysis of the relationships among brain size, cell number, connectivity, and regional specialization. If cell number increases with brain size, the amount of fiber needed to maintain the same connectivity increases disproportionately. The result is that larger brains will tend to consist of a larger number of specialized regions than will smaller brains, where broad interconnectivity can more easily be achieved.

It is important to note that adaptation and constraints may both play a role in the evolution of a feature, and these hypotheses are not mutually exclusive. For example, selection might favor an increase in brain size; because of constraints on connectivity, the result may be a larger brain (adaptation) with an increased number of discrete subdivisions (constraint). Many features that have arisen through convergent evolution might represent a combination of adaptation and constraints.

Organization of Olfactory Systems

In general, the term *olfaction* is applied to chemosensory systems that detect chemicals emanating from a distant source. Other chemosensory systems generally require physical contact with the source for detection, and in invertebrates such systems are called 'gustatory'. Vertebrates have anatomically distinct olfactory and gustatory systems, as well as other chemoreceptors, such as the common chemical sense and solitary chemoreceptors. Because analogous systems in other phyla are difficult to identify, I will not consider these chemosensory systems here. In this paper, I will also not consider examples presented by the vomeronasal system, a vertebrate olfactory subsystem frequently but erroneously assumed to be convergent with specialized pheromone-sensing systems of insects [Eisthen, 1997].

The olfactory systems of vertebrates, molluscs, arthropods, and nematodes share many features that are intriguingly similar, and markedly different from the features of other sensory systems. In general, olfactory receptor cells are bipolar neurons with a dendrite that protrudes into a fluid medium. The dendrite is capped with cilia and/or microvilli, which are probably the site of odorant transduction, and where the membrane-bound odorant receptors are presumably localized. The odorant receptor genes are members of a large superfamily that produces molecules with seven membrane-spanning regions; most of these receptors interact with G proteins that activate intracellular signaling pathways. At the opposite pole of the cell, the axon projects directly into the central nervous system, usually without branching, and often forms synapses with many other cells in tangles of fibers known as *glomeruli*.

Table 1. Model organisms commonly used in neurobiological olfactory research

Phylum	Class	Genus and species	Common name(s)
Nematodes	Secernenteans	***Caenorhabditis elegans***	
Arthropods	Malacostracans	*Homarus americanus*	American lobster
		Panulirus argus	Caribbean spiny lobster
	Insects*	*Periplaneta americana*	American cockroach
		Apis mellifera	honeybee
		Manduca sexta	tobacco hornworm, hawk moth, sphinx moth
		Heliothis virescens	tobacco budworm
		Schistocerca americana	American grasshopper
		Drosophila melanogaster	fruit fly
Molluscs	Gastropods	*Achatina fulica*	giant African snail
		Helix aspersa	brown garden snail
		Limax (Lehmannia) marginatus	tree slug
		Limax maximus	spotted garden slug, great gray garden slug, European giant garden slug
Vertebrates	Actinopterygians	***Brachydanio rerio***	zebrafish
		Carassius auratus	goldfish
		Ictalurus punctatus	channel catfish
		Oncorhynchus mykiss	rainbow trout
	Amphibians	***Ambystoma tigrinum***	tiger salamander
		Necturus maculosus	mudpuppy
		Rana spp.**	true frogs
		Xenopus laevis	African clawed frog
	Mammals	*Mesocricetus auratus*	Syrian hamster, golden hamster
		Mus musculus domesticus	house mouse
		Rattus norvegicus	Norway rat
		Cavia porcellus	guinea pig
		Sus scrofa	pig
		Bos taurus	cow
		Ovis aries	sheep
		Homo sapiens	human

Animals that have proved to be particularly valuable models, for which much data are available, are indicated in ***boldface.***

* Many insects, particularly those of agricultural importance, have been studied; most such research focuses more on chemical ecology and behavior than on neurobiology.

** Various species of ranid frogs are used in olfactory research, including *Rana catesbeiana, R. pipiens, R. ridibunda, R. temporaria,* and *R. 'esculenta'* [a naturally-occurring hybrid of *R. ridibunda* and *R. lessonae*; Berger, 1967, 1968].

The olfactory systems of vertebrates possess all these features, and those of animals in other phyla generally conform to this pattern, with a few exceptions. Vertebrate olfactory receptor neurons are found in a pseudostratified epithelium inside the nasal cavity, and the axons project to glomerular structures in the olfactory bulb at the rostral end of the telencephalon. In insects, the olfactory receptor neurons are found in clusters inside cuticle-covered sensillae along the antennae, and sometimes on other appendages such as the maxillary palp in *Drosophila* [Singh and Nayak, 1985; Ayer and Carlson, 1992]. Odorants gain access to olfactory receptor neurons via pores in the sensillar cuticle. The axons of the olfactory receptor neurons project to the antennal lobe in the brain. In lobsters, olfactory receptor neurons are found in clusters inside specialized sensillae along the antennules, and the axons of these cells project to the olfactory lobe in the deutocerebrum. The olfactory receptor neurons of snails are clustered in groups, and their dendrites project to a dense olfactory epithelium at the tip of the tentacle. The axons of most of these cells project a short distance into regions associated with the tentacle ganglion [Chase and Tolloczko, 1993]. In *Caenorhabditis elegans*, olfactory receptor neurons are found in the paired amphid organs near the mouth, which are sensitive to mechanosensory stimuli as well as chemicals in water (gustation) and volatile chemicals in air (olfaction). Of the 12 neurons in each amphid organ, 3 respond exclusively to volatile compounds and may be considered purely olfactory neurons, and another 2 respond to volatile chemicals as well as other stimuli [Bargmann et al., 1993; Mori and Ohshima, 1997]. The cell bodies of these neurons are located in the lateral ganglia, and make synapses with each other and with interneurons in the lateral and ventral ganglia [White et al., 1986].

Our ability to recognize cases of similarity and analyze underlying causes is necessarily limited by the available data. During the past 40 years, the most popular model animals for laboratory studies of the anatomy and physiology of the olfactory system have been a few selected species of insects, lobsters, snails, fishes, salamanders, and rodents. More recently, researchers have initiated studies of chemoreception in the nematode *C. elegans*, and have made such progress that we can reasonably compare the organization of the olfactory system of *C. elegans* with those in members of other phyla. Table 1 contains a list of animals for which a substantial body of data is available concerning the organization and function of the peripheral olfactory system and/or the first central target. In analyzing the features of olfactory systems in diverse animals, we must bear in mind the problem of incomplete data sets, as the information available for a given species reflects the strengths of that animal as a laboratory model. For example, the sphinx moth *Manduca sexta* has proved to be an excellent model for anatomical and electrophysiological studies because of its large brain and large, accessible antennae; in contrast, the fruit fly *Drosophila melanogaster* is more suitable for molecular and genetic studies, but because of its small size fewer anatomical and electrophysiological experiments have been carried out.

In this paper, I will discuss four features that are found in olfactory systems in more than one phylum. These features are: (1) the presence of odorant binding proteins in the fluid overlying the receptor cell dendrite; (2) the use of G protein-coupled receptors as odorant receptors; (3) the use of a two-step signaling cascade in odorant transduction; and (4) the presence of glomerular structures at the first central target in the olfactory pathway. These features may represent adaptations that have evolved independently, and therefore might provide us with valuable information about the way the nervous system processes odorant stimuli. Alternatively, these similar features may instead reflect underlying homology, or could have arisen independently due to similar constraints. A further possibility should be considered the null hypothesis: the perceived similarity may be superficial, and may not reflect any of these processes. For each olfactory system feature, I will consider the data bearing on each of these hypotheses to try to determine whether or not the feature has arisen as a result of convergence; if so, I will assess the evidence that the feature arose as an adaptation or as a result of constraints. I will then discuss the relevance of this analysis for our understanding of olfactory system function, and for the use of evolutionary convergence as a tool in neurobiology. In the following analysis, I will generally assume that features present in several members of a large group are homologous within that group, although there is clear evidence that characters can evolve independently several times even within narrow taxonomic groups [for examples, see Bell, 2002; Nishikawa, 2002; Wray, 2002].

Odorant Binding Proteins

Before an odorant molecule can bind to a receptor, it must pass through the fluid overlying the olfactory receptor neurons, above the sensory epithelium or within the sensillum or amphid organ. In some insects and mammals, this fluid has been found to contain specialized molecules that bind odorants, and which are called odorant binding proteins (OBPs) [Vogt and Riddiford, 1981; Pelosi et al., 1982; Pevsner et al., 1985]. OBPs are soluble proteins dissolved in the fluid overlying the receptor neurons, and are not membrane-bound.

The first OBP was discovered in the pheromone-specific sensillae of the silk moth *Antheraea polyphemus* [Vogt and Riddiford, 1981]. OBPs have since been found in at least 17 species of endopterygote insects and 1 species of hemipteran insect [Vogt et al., 1999]. Within species, multiple OBPs have been identified. For example, 7 different OBPs have been isolated from *Manduca*, and 17 have been found in *Drosophila* [Robertson et al., 1999]. The molecules that function as OBPs in insects are of similar size and are characterized by six cysteines that are found in particular portions of the sequence, but these molecules also have diverse structures, even within a species, and do not appear to belong to a single class of proteins [Vogt et al., 1999]. Each type of OBP appears to have a fairly narrow odorant binding affinity and is found in the lymph of particular subsets of sensillae, rather than being broadly distributed across the antenna [Vogt et al., 1991]. Within a single sensillum, more than one OBP may be expressed [Hekmat-Scafe et al., 1997].

In mammals, all OBPs discovered to date are lipocalins, and are therefore structurally less diverse than those found in insects. The lipocalin family comprises a group of carrier proteins with diverse functions, including retinol binding proteins, α_2 microglobulins, and the major urinary proteins (MUPs) [Tegoni et al., 2000]. Interestingly, MUPs have been shown to bind with pheromones in urine [Singer, 1991; Bacchini et al., 1992; Böcskei et al., 1992; Robertson et al., 1993], and may also serve as pheromones themselves [Singer and Macrides, 1993; Mucignat-Caretta et al., 1995]. As in insects, the olfactory mucus contains several different OBPs. The largest number found to date is in porcupines *(Hystrix cristata)*, which have at least 8 different OBPs [Felicioli et al., 1993]. Mice *(Mus musculus domesticus)* have 6 different OBPs that are not closely related within the lipocalin family [Utsumi et al., 1999]. Multiple OBPs have also been found in rabbits [probably *Oryctolagus cuniculus;* Garibotti et al., 1997], rats [*Rattus norvegicus;* Löbel et al., 2001], and humans [Lacazette et al., 2000]. Mammalian OBPs, like those of insects, bind ligands selectively [Löbel et al., 2002]; however, unlike those of insects, mammalian OBPs are not restricted to the fluid near particular olfactory receptor cells, and are distributed throughout the mucus overlying the olfactory epithelium, as well as adjacent regions of nonsensory epithelium.

An olfactory-specific lipocalin that might function as an OBP has been described in the clawed frog *Xenopus laevis* [Lee et al., 1987], but Baldaccini et al. [1986] were unable to measure OBP-like binding in another amphibian, the common frog *(Rana temporaria)*. Indeed, these authors examined OBP-like activity in a variety of vertebrates, and were able to demonstrate binding in several mammalian species but not in birds (rock doves, *Columba livia,* and Muscovy ducks, *Cairina moschata*), Hermann's tortoises *(Testudo hermanni),* teleost fishes (American eels, *Anguilla rostrata,* rainbow trout, *Salmo gairdneri,* and black bullhead catfish, *Ictalurus melas*), or thornback rays *(Raja clavata)*. OBPs have not been reported in other non-mammalian vertebrates, in lobsters, or in *C. elegans*. OBPs have not been identified in molluscs, but Chase and Tolloczko speculate that a protein they found in the mucus overlying the olfactory epithelium in snails *(Achatina fulica)* may function as an OBP [Chase and Tolloczko, 1993]. The phylogenetic distribution of known OBPs is illustrated in figure 1.

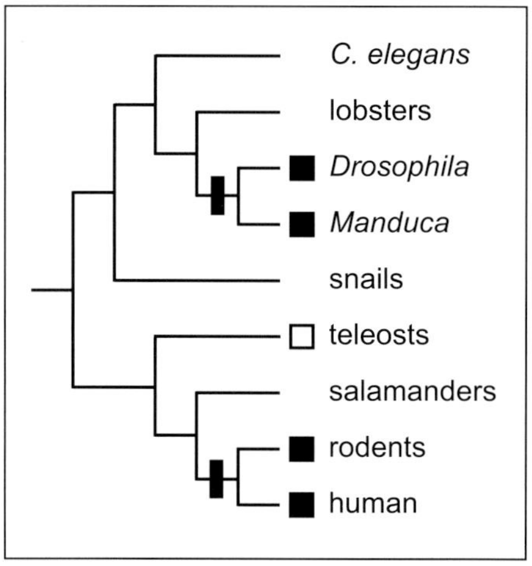

Fig. 1. Phylogenetic distribution of odorant binding proteins (OBPs) among common model animals used in olfactory research. Taxa in which OBPs have been identified are indicated with solid boxes, and animals for which OBPs may not be present are indicated with an empty box. Based on these data, a hypothesis concerning the evolutionary history of OBPs is illustrated, with separate origins indicated by bars; the hypothesis illustrated here postulates that OBPs evolved once in terrestrial insects and once separately in mammals.

Why do we find OBPs in insects and mammals? Is the presence of OBPs in these two groups an example of evolutionary convergence that informs us about the workings of the olfactory system, or are alternative hypotheses also tenable?

Homology

The presence of OBPs in insects and mammals could be due to homology; that is, OBPs could have arisen once, in the common ancestor of insects and mammals. Given the complete lack of similarity in the structure of mammalian and insect OBPs, we can reject this hypothesis, and conclude that OBPs have arisen independently at least twice.

Convergence

Given that OBPs have arisen independently in insects and vertebrates, they may be an example of convergent evolution. If so, are they adaptations that serve a particular function, or did they arise as the result of constraints?

Adaptation. Several hypotheses about the adaptive function of OBPs have been proposed, and fall into three categories. All draw support from studies demonstrating that OBPs in insects and mammals bind odorants selectively, and in some insects are associated with functionally specialized sensillae [Vogt et al., 1991; Du and Prestwich, 1995; Steinbrecht, 1996; Plettner et al., 2000; Löbel et al., 2002]. (1) OBPs may serve as filters before odorant stimuli arrive at the receptors [Vogt et al., 1991]. (2) OBPs form a complex with odorants, and the complex may interact with the odorant receptor [Prestwich et al., 1995]. (3) OBPs might inactivate odorants, perhaps to prevent desensitization and/or to allow detection of new stimuli. Different mechanisms of inactivation have been proposed, including the possibility that OBPs directly inactivate odorants by binding them [Vogt and Riddiford, 1981], that the OBP-odorant complex serves as a substrate for enzymes that degrade odorants [Vogt et al., 1985], or that OBPs remove excess odorant from the lymph or mucus surrounding the receptor neurons [Kaissling, 1998].

Constraint. Similar features can arise independently because of constraints. Although it is difficult to imagine how the existence of OBPs could result from developmental or genetic constraints, OBPs might have arisen in terrestrial insects and mammals because of physical constraints imposed by the problem of detecting odorants in air. Specifically, the observation that OBPs have been identified in terrestrial animals but not in any aquatic animal has led to the popular hypothesis that OBPs function in insects and mammals to transport hydrophobic molecules through lymph or mucus to the odorant receptors [Bignetti et al., 1987; Vogt, 1987]. If so, it is difficult to understand why OBPs have not been found in non-mammalian terrestrial vertebrates [Baldaccini et al., 1986], but perhaps broader assays would detect different lipocalins or other classes of odorant-binding molecules in these animals. In any case, this hypothesis could be falsified independently for insects and for mammals by the discovery of homologous proteins in the olfactory sensillae or epithelia of aquatic relatives. Conversely, this hypothesis would gain support if researchers could demonstrate that OBPs bind to a membrane-bound docking protein, releasing odorant very close to the receptor, as has been suggested by Rogers and colleagues [Rogers et al., 1997]. Indeed, a membrane-bound OBP receptor has been discovered in cows [Boudjelal et al., 1996], although its relationship with odorant receptors is still unclear.

The Null Hypothesis

In considering a feature that has arisen independently several times, the null hypothesis is that the observed similarities are superficial and of no functional or evolutionary significance. Perhaps the presence of proteins that can bind odorants in sensillar lymph and olfactory epithelial mucus is coincidental, and OBPs do not serve the same function in insects and mammals. A variant null hypothesis suggests that OBPs may serve a general role not specific to olfaction, such as detoxification of external chemicals [Boudjelal et al., 1996]. Support for the latter hypothesis comes from the observation that, in mammals, OBPs and other lipocalins are distributed throughout the mucus overlying the nasal epithelia, including nonsensory regions, and that OBP-like lipocalins are found in non-nasal tissues [Lacazette et al., 2000]. Similarly, in insects, homologues of OBPs have been found in non-olfactory tissues [Vogt et al., 1999]. Because the function of OBPs has not yet been demonstrated for any species, we cannot discount either of these null hypotheses.

Although we cannot at present dismiss the null hypotheses, OBPs may represent an example of convergent evolution. If so, the evolution of OBPs may have been shaped by both adaptation and constraints. Indeed, features that arise to serve one function are not infrequently co-opted for another function later [Gould and Vrba, 1982]. For example, OBPs might have initially arisen to transport odorants across the air/liquid interface surrounding receptor neurons (i.e., as a response to a constraint), and each organism originally had only one form of OBP; the OBP genes may then have duplicated and diverged, and different OBPs began to function in filtering odorant stimuli before they arrive at the receptors. Conversely, OBPs may have arisen to function in odorant processing (i.e., as an adaptation), and their *absence* in aquatic organisms may reflect a constraint. OBPs are present in mucus or lymph at concentrations up to 10 mM and are water-soluble [Vogt and Riddiford, 1981]. Perhaps OBPs serve a valuable function in odorant processing in insects and mammals, and aquatic animals such as fishes are constrained from using OBPs because of the high energetic cost of producing proteins that could diffuse away into the environment.

These different scenarios can be supported or refuted by new data. If further research reveals that OBPs serve the same function in insects and mammals and are not present in aquatic animals, we may conclude that OBPs are an adaptation and are constrained from being used by aquatic animals. On the other hand, if we learn that OBPs are only present in terrestrial vertebrates and insects but serve different functions in these groups, we might conclude that OBPs arose to mitigate the constraints involved in detecting hydrophobic odorants in air, and that they were secondarily co-opted for different uses in odorant processing. Finally, if the distribution of OBPs is not confined to terrestrial animals and OBPs serve different functions in different groups, or serve a general function not specific to olfaction, we may conclude that any perceived similarities were superficial, and that the null hypothesis is supported.

Structure of Odorant Receptors

Odorant receptors are part of the large, diverse superfamily of G protein-coupled receptors (GPCRs) with seven membrane-spanning domains. The family of GPCRs includes opsins and muscarinic acetylcholine receptors as well as receptors for sweet taste, serotonin, dopamine, prostaglandin, and gonadotropin releasing hormone. Odorant receptors were first described in rats *(R. norvegicus)* by Buck and Axel [1991], who developed probes for GPCRs based in part on physiological and biochemical evidence indicating that odorant binding activates G proteins. Homologous odorant receptor genes have now been found in more than 20 mammalian species, as well as in birds, amphibians, coelacanths, teleosts, and lampreys [Ngai et al., 1993a; Freitag et al., 1995; Nef et al., 1996; Freitag et al., 1998, 1999]. The vertebrate odorant receptor genes constitute large families: catfish *(I. punctatus)* and zebrafish *(Brachydanio rerio)* are estimated to have about 100, and mice *(M. musculus domesticus)* have at least 100 and perhaps as many as 1000 odorant receptor genes [Buck and Axel, 1991; Levy et al., 1991; Ngai et al., 1993a; Barth et al., 1996].

The identification of odorant receptors in non-vertebrates has been difficult. Nevertheless, odorant receptor genes were recently sequenced from *Drosophila* and *C. elegans* using a bioinformatics approach, in which researchers examined data derived from genome projects for sequences encoding proteins that would be predicted to have seven transmembrane domains [Troemel et al., 1995; Clyne et al., 1999; Gao and Chess, 1999; Vosshall et al., 1999]. Although it is conceivable that insects, nematodes, and vertebrates could use fundamentally different receptors for odorant detection, the large numbers of GPCR-type odorant receptor genes found seem to rule this out: *Drosophila* has about 60 odorant receptor genes and *C. elegans* may have 100 – 500 [Troemel et al., 1995; Bargmann, 1998; Vosshall et al., 2000]. Nevertheless, the sequences of the odorant receptor genes in *C. elegans* and *Drosophila* are quite different from those of vertebrates, indicating that the odorant receptor genes in these three phyla were co-opted independently from the larger family of GPCR genes. The phylogenetic distribution of known G protein-coupled odorant receptors is illustrated in figure 2.

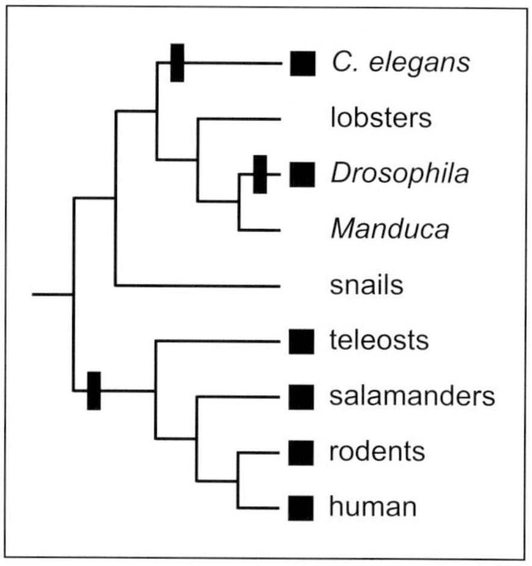

Fig. 2. Phylogenetic distribution of G protein-coupled odorant receptors among common model animals used in olfactory research. Taxa in which these receptors have been identified are indicated with solid boxes. A hypothesis concerning the evolutionary history of these receptors is illustrated. Given that the receptor genes are quite different in *C. elegans, Drosophila,* and vertebrates, the hypothesis illustrated here postulates that members of the large family of GPCR genes were co-opted independently at least three times to serve as odorant receptors.

Some features of the odorant receptor genes are similar across phyla, and some differ. The number of receptor genes expressed in each olfactory neuron remains controversial, because direct measures are extremely difficult to obtain. Based on the proportion of olfactory neurons hybridizing with individual odorant receptor gene probes, evidence that individual olfactory receptor neurons express a single allele of an odorant receptor gene, and the observation that neurons expressing the same receptor project to the same glomerulus, several groups have proposed that each vertebrate olfactory receptor neuron expresses only one odorant receptor gene [Nef et al., 1992; Ngai et al., 1993b; Chess et al., 1994; Ressler et al., 1994; Vassar et al., 1994]. Contrary evidence has been obtained from goldfish *(Carassius auratus)*, in which two odorant receptor genes are each expressed in a large proportion of the olfactory receptor neurons, suggesting that each neuron expresses at least 2 or 3 receptor genes [Speca et al., 1999]. In *Drosophila,* each olfactory receptor neuron expresses the OR83b receptor gene and at least one other, but probably not more than a small number of receptor genes [Vosshall et al., 1999, 2000]. In contrast, *C. elegans* olfactory receptor neurons may express many odorant receptor genes. Troemel et al. conservatively estimate the total number of *C. elegans* odorant receptor genes at 100, but a more recent estimate based on the full genome suggests that the actual number could be closer to 500 [Troemel et al., 1995; Bargmann, 1998]. Given that *C. elegans* has 10 neurons that respond to volatile odorants [Bargmann et al., 1993; Mori and Ohshima, 1997], each olfactory neuron may express 10–50 odorant receptor genes, far more than has been proposed for vertebrates or *Drosophila.*

Odorant receptor genes are expressed in zones in both vertebrates and *Drosophila.* In mammals, odorant receptor genes are expressed in one of four broad zones in the olfactory epithelium, and expression of a single gene is scattered randomly within a particular zone [Ressler et al., 1993; Vassar et al., 1993; Strotmann et al., 1994]. In the olfactory rosette of zebrafish, each receptor gene is expressed in one of three concentric, slightly overlapping zones [Weth et al., 1996]. The odorant receptor genes in *Drosophila* are expressed in a variable number of sensillae in a bilaterally symmetrical pattern on the antennae. The number and position of the sensillae in which a receptor is expressed appear to be consistent across individuals [Vosshall et al., 2000]. Again, the olfactory system of *C. elegans* appears to be organized differently, as there is no evidence of a zonal pattern of expression of odorant receptor genes [Troemel et al., 1995].

The ligand specificity of the odorant receptors may be similar across phyla. The first demonstration of the function of a putative odorant receptor was accomplished in *C. elegans,* using a behavioral assay of odorant sensitivity in animals with mutations in the *odr-10* gene. Of the seven odorants tested, the animals failed to respond only to diacetyl, suggesting that the gene codes for a very narrowly-tuned odorant receptor [Sengupta et al., 1996]. When expressed in a human cell line, however, the ODR-10 receptor responded to diacetyl and to one of several structurally-related compounds that were tested, as well as to one structurally dissimilar chemical [Zhang et al., 1997]. Thus, the ODR-10 receptor is fairly narrowly tuned, but can interact with more than one ligand. In rats, when the I7 odorant receptor gene was overexpressed in the olfactory epithelium, the response to octyl aldehyde was greatly enhanced, responses to the related compounds heptaldehyde, nonyl aldehyde, and decyl aldehyde were increased, and responses to 70 other odorants were unchanged [Zhao et al., 1998]. Studies of mammalian odorant receptors in heterologous expression systems have found that the cells respond to a subset of the odorants tested, and show some selectivity for families of structurally similar compounds [Krautwurst et al., 1998; Murrell and Hunter, 1999]. A goldfish odorant receptor expressed in *Xenopus* oocytes responded to basic and neutral aliphatic L-amino acids, with a high affinity for arginine and lysine, and did not respond to 13 other amino acids, 10 amino acid derivatives and neurotransmitters, or to other behaviorally important odorants such as bile acids, prostaglandins, and sex steroids [Speca et al., 1999]. Relatively broad ligand binding has also been suggested by studies in which the *Drosophila* Or43a receptor gene was overexpressed in antennal olfactory receptor neurons or expressed in *Xenopus* oocytes [Störtkuhl and Kettler, 2001]. Based on the data available, it appears that odorant receptors in *C. elegans, Drosophila,* and vertebrates respond to multiple odorants, and in some cases appear to respond preferentially to groups of structurally similar compounds.

Why do we find G protein-coupled odorant receptors in *C. elegans, Drosophila,* and vertebrates? In principle, other types of receptors, such as ligand-gated ion channels, could be used as odorant receptors. Is the presence of G protein-coupled odorant receptors in these phyla an example of evolutionary convergence that informs us about the workings of the olfactory system, or are alternative hypotheses also tenable?

Homology

GPCRs may have been co-opted once for use in odorant detection, in which case the odorant receptors in *C. elegans, Drosophila,* and vertebrates would be descended from, and therefore homologous with, these original odorant receptors. Given the very low level of sequence similarity among odorant receptor gene families in these different phyla, we can rule out this possibility, and conclude that the odorant receptors were co-opted independently out of the larger family of GPCRs. Furthermore, the status of GPCRs as a monophyletic family is questionable [Josefsson, 1999].

Convergence

If odorant receptors have been co-opted from GPCR families that arose independently, they could not be considered homologous, but are examples of structural convergence [Doolittle, 1994; Zakon, 2002]. Are they adaptations, or responses to constraints?

Adaptation. The occurrence of G protein-coupled odorant receptors in nematodes, insects, and vertebrates may constitute independent adaptations for transducing odorant stimuli. If so, why? Is there something about these receptors that makes them particularly suitable for use in olfaction? Based on a consideration of the properties of GPCRs, three explanations seem plausible. These hypotheses are not mutually exclusive, and all these properties of GPCRs could have contributed to their adoption as odorant receptors. (1) The use of GPCRs allows for amplification of small signals [Selbie and Hill, 1998], a feature that would be adaptive in situations in which few odorant molecules reach the olfactory epithelium. Although the lower limit of detectability has been debated, some studies have suggested that olfactory receptor neurons in both insects and vertebrates may be able to respond to single odorant molecules [Kaissling, 1986; Menini et al., 1995]. (2) Another potential functional advantage to the use of GPCRs is that different odorants could bind to a single receptor type and activate different, interacting signaling pathways through one or more G proteins [Selbie and Hill, 1998]. Such flexibility may provide individual olfactory receptor neurons with the ability to respond to different odorants in different ways, even if each neuron expresses only a single receptor gene. (3) GPCRs can form homo- and heterodimers [Bouvier, 2001], and recent studies suggest that G protein-coupled sweet taste receptors form dimers [Max et al., 2001; Nelson et al., 2001]. If odorant receptors can also form heterodimers, the number of combinations that could be formed from the large families of odorant receptor genes is virtually limitless. The ability of GPCRs to form dimers may make them particularly suitable for transducing or coding the broad array of odorants that many animals encounter.

Constraint. Perhaps the repeated deployment of GPCRs for use in olfaction reflects the action of a constraint, such as a developmental or genetic constraint. For example, perhaps members of this gene family are particularly easily duplicated, and this creates a bias such that these receptor genes are more likely than others to be coopted for novel functions. I know of no data that would support such a scenario, but it cannot be dismissed at present.

The Null Hypothesis

Finally, we must consider the possibility that the discovery of G protein-coupled odorant receptors in vertebrates, *Drosophila,* and *C. elegans* is misleading. One could argue that because researchers found G protein-coupled odorant receptors in *Drosophila* and *C. elegans* only by combing through genomic data seeking receptors similar to those found in vertebrates, the results are biased, the similarity is superficial, and that other receptor types also serve as odorant receptors in these animals but have not been found because we have not looked for them. This position seems untenable: the large numbers of G protein-coupled odorant receptor genes found in *Drosophila* and *C. elegans* suggest that these genes code for all or almost all the odorant receptors in these animals; further, members of both gene families have been demonstrated to function in odorant transduction, as described above.

In summary, the use of GPCRs as odorant receptors in *C. elegans, Drosophila,* and vertebrates appears to be an example of evolutionary convergence, and the use of these receptors may constitute an adaptation for the processing of odorant signals, although the functional advantages of their use are not yet clear. GPCRs seem to be evolutionarily and physiologically flexible, as they can interact with each other and with a variety of different intracellular signaling pathways.

Signal Transduction Strategies

The binding of odorants to receptors can cause olfactory receptor neurons to depolarize or hyperpolarize, and can cause changes in baseline membrane conductance. These effects are mediated by a variety of different signal transduction pathways, which might co-exist within individual cells, in teleosts, amphibians, rodents, lobsters, squid, and *C. elegans* [Schild and Restrepo, 1998; Ache and Restrepo, 2000]. In this paper, I will concentrate on one particular olfactory transduction strategy, as it is unusual among sensory systems and has been well documented in olfactory receptor neurons in two phyla: receptor binding initiates a two-step signal transduction cascade in which cation channels open, and the cations that enter the cell then gate additional ion channels, contributing to depolarization of the cell [Ache and Restrepo, 2000].

A two-step signal transduction pathway has been described in detail in rats *(R. norvegicus)* and mice *(M. musculus domesticus)* in electrophysiological, biochemical, and molecular/genetic studies. Odorant binding activates an olfactory-specific G protein, stimulating type III adenylyl cyclase, which increases cAMP levels inside olfactory receptor neurons [Sklar et al., 1986; Jones and Reed, 1989; Bakalyar and Reed, 1990; Belluscio et al., 1998; Wong et al., 2000]. cAMP gates a non-selective cation conductance that is permeable mainly to calcium, and the calcium that enters through these channels gates calcium-dependent chloride channels [Dhallen et al., 1990; Brunet et al., 1996]. Because intracellular chloride levels are relatively high in olfactory receptor neurons, Cl^- flows outward, depolarizing the cell [Lowe and Gold, 1993; Reuter et al., 1998]. Components of this pathway have been described in humans and in cows *(Bos taurus),* suggesting that this transduction strategy might be common among mammals [Ludwig et al., 1990; Gomez et al., 2000].

In other classes of vertebrates, this two-step transduction pathway appears to be present, although it has not been characterized as fully as in rodents. For example, electrophysiological experiments demonstrate that these signaling mechanisms are also active in olfactory receptor neurons of salamanders *(Ambystoma tigrinum, Cynops pyrrhogaster,* and *Necturus maculosus)* and anurans *(Bufo marinus, Rana catesbeiana, R. pipiens, R. 'esculenta', R. ridibunda,* and *Xenopus laevis)* [Sklar et al., 1986; Nakamura and Gold, 1987; Firestein et al., 1991; Frings and Lindemann, 1991; Kleene and Gesteland, 1991; Dubin and Dionne, 1993; Kurahashi and Yau, 1993; Zhainazarov and Ache, 1995a]. Studies with olfactory receptor neurons from catfish *(Ictalurus punctatus),* zebrafish *(B. rerio),* and carp *(Cyprinus carpio)* demonstrate that a cyclic nucleotide-gated cation channel is present, that odorant binding elevates cAMP levels, and that cAMP activates a cation current [Goulding et al., 1992; Kolesnikov and Kosolapov, 1993; Ma and Michel, 1998], indicating that the first step in a two-step transduction cascade is present. The second step might also be present in teleosts: in rainbow trout *(Oncorhynchus mykiss)* a calcium-dependent chloride conductance is activated by odorants [Sato and Suzuki, 2000]. Taken together, these data indicate that the two-step signal transduction pathway that has been described in rodents is also present in amphibians and teleost fishes, suggesting that it is widely used in vertebrates.

Studies from Ache's laboratory have elucidated a different two-step transduction pathway that is activated by odorant binding in Caribbean spiny lobsters *(Panulirus argus).* Although the links among steps in this pathway are not as firmly established as in rodents, odorants have been demonstrated to activate a depolarizing cation current in spiny lobster olfactory receptor neurons [Anderson and Ache, 1985]. The underlying mechanism involves odorant activation of a G_q, a G protein that is associated with phospholipase C [Fadool et al., 1995]. The second messenger that activates the cation current seems to be IP_3, as odorant exposure elevates IP_3 levels, receptors for IP_3 are found in the ciliary membrane, and IP_3 activates a depolarizing cation current [Fadool and Ache, 1992; Boekhoff et al., 1994; Hatt and Ache, 1994; Munger et al., 2000; Zhainazarov et al., 2001]. Thus, although the second messenger in the first step of the transduction pathway in spiny lobsters is IP_3, rather than cAMP as in vertebrates, the result is the same: odorant exposure causes a cation channel to be opened. Interestingly, sodium-dependent non-selective cation channels are also present in outer dendrites of spiny lobster olfactory receptor neurons. Depolarizing odorant responses are reduced by drugs that block the sodium-dependent cation channel and by the substitution of other cations for sodium in the extracellular fluid, suggesting that the sodium-dependent cation channel in spiny lobster olfactory receptor neurons plays a role analogous to that of the calcium-activated chloride channel in vertebrate olfactory receptor neurons [Zhainazarov and Ache, 1995b, 1997; Zhainazarov et al., 1998]. Thus, it appears that in spiny lobsters, odorant binding elevates IP_3 levels, gating a cation channel, and the sodium that enters the cell through this cation channel gates a second, non-selective cation channel.

Can we find components of a two-step transduction pathway in other groups of animals? The pathways used in both vertebrates and spiny lobsters possess some common features: odorant binding increases levels of a second messenger that gates a cation channel, and the entering cations are involved in gating an additional channel. In *C. elegans,* the first step in such a pathway appears to be present. The *C. elegans* genome project has revealed the existence of 20 genes coding for alpha subunits of G proteins, two of which are necessary for chemotaxis to some volatile odorants [Jansen et al., 1999]. Cyclic nucleotide-gated cation channels are present in a subset of olfactory receptor neurons, are gated by cGMP, and are necessary for chemotaxis [Coburn and Bargmann, 1996; Komatsu et al., 1996, 1999]. Data concerning a second step in the transduction pathway, such as an additional cation-activated channel, are lacking.

In insects, a depolarizing odorant-induced current appears to be mediated by G protein-activated stimulation of IP_3, as has been demonstrated in American cockroaches *(Periplaneta americana),* migratory locusts *(Locusta migratoria),* sphinx moths *(M. sexta),* and fruit flies *(D. melanogaster)* [Boekhoff et al., 1990a, b; Breer et al., 1990; Stengl, 1994; Riesgo-Escovar et al., 1995]. Stengl's description of odorant-induced currents in cultured *Manduca* olfactory receptor neurons suggests the presence of a two-step transduction process: odorants increase levels of IP_3, which stimulates an IP_3-dependent calcium current; the influx of calcium then stimulates a calcium-dependent cation current [Stengl, 1994]. Although the details of the channels and second messengers involved are not known, these data hint that different two-step processes might mediate odorant transduction in vertebrates, spiny lobsters, and *Manduca.*

Figure 3 illustrates the phylogenetic distribution of animals in which the presence of a two-step odorant transduction pathway is indicated. A transduction strategy that involves the use of external cations to gate additional ion channels is unusual in sensory receptor cells. Why do we find this transduction strategy in the olfactory systems of spiny lobsters, *Manduca,* and vertebrates? Is this an adaptation for processing odorant stimuli, or should we consider other hypotheses as well?

Homology

In theory, the presence of a two-step transduction strategy in these groups could be due to inheritance from a common ancestor, and represent an example of homology. This hypothesis is simply not supportable: the components of the pathway differ greatly among *Manduca,* spiny lobsters, and vertebrates, including the type of G-protein, the second messenger used, the primary cation channel, and the second ion channel that is activated.

Convergence

The separate origins of two-step transduction pathways may be due to evolutionary convergence, and might constitute an adaptive mechanism that helps us understand the ways in which olfactory information is processed by the nervous system.

Adaptation. Several hypotheses concerning the adaptive function of a two-step transduction strategy have been proposed. (1) The use of a second ion channel that opens as an indirect result of odorant binding may serve to amplify the signal. This feature would be particularly useful in a sensory system such as olfaction that detects low-concentration or infrequent stimuli [Kleene, 1993; Lowe and Gold, 1993; Zhainazarov and Ache, 1995b]. (2) A variant on this hypothesis states that the use of a second ion channel might allow for *regulation* of amplification of the signal. For example, the degree of amplification achieved by opening chloride channels in vertebrate olfactory receptor neurons could be regulated by altering the sensitivity of the channel to calcium, or by regulating the concentration of calcium inside the cell [Ache and Restrepo, 2000]. (3) The contribution of the chloride current in vertebrate olfactory receptor neurons depends on the internal concentration of chloride. If the concentration is not as high as some have suggested [Reuter et al., 1998], then rather than contributing to depolarization, the chloride current might limit depolarization or contribute to the repolarization of the cell [Kleene and Gesteland, 1991; Kleene, 1993]. This argument seems to apply only to the transduction pathway described in vertebrates, as the ion channels involved in transduction in spiny lobsters and *Manduca* pass external cations.

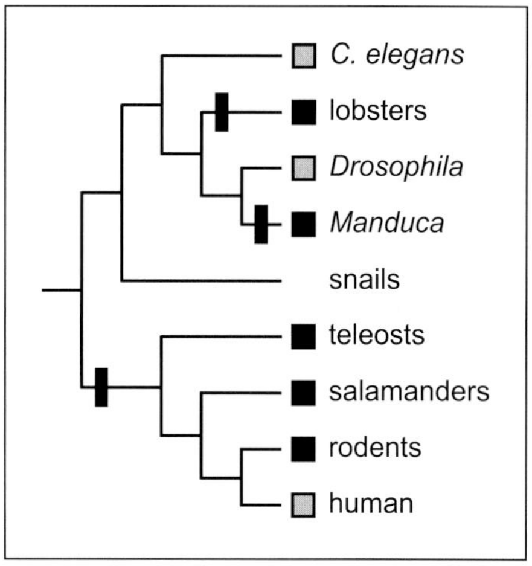

Fig. 3. Phylogenetic distribution of two-step odorant transduction processes among common model animals used in olfactory research. Taxa in which a two-step transduction process has been demonstrated are indicated with solid boxes; gray boxes indicate taxa for which odorant-gated cation currents or cation channels, the first step in such pathways, have been described. Based on these data, a hypothesis concerning the evolutionary history of this transduction strategy is illustrated. Although data from suitable outgroups are lacking, the components of the pathways differ greatly among groups, suggesting that this transduction strategy evolved independently in the ancestors of vertebrates, lobsters, and *Manduca.* The strategy used by *Manduca* and *Drosophila* may be the same, but the available data are too sparse for hypotheses to be proposed.

Constraint. Two-step transduction pathways might have arisen because of a physical constraint that is unique to olfactory receptor neurons. Indeed, such a hypothesis has been proposed to explain the use of a secondary step involving chloride ions in vertebrate olfactory receptor neurons. If the ionic composition of the mucus layer is difficult to regulate, the ionic environment of olfactory receptor neurons could be somewhat unpredictable, particularly for aquatic organisms. Thus, it is not always certain that enough external cations will be available to depolarize the cell; however, depolarization can be ensured through the use of a step that involves the efflux of anions [Kurahashi and Yau, 1993]. This hypothesis is attractive, but only applies when the second ion channel in the transduction pathway passes internal anions. That is, the hypothesis might explain the evolution of a two-step transduction pathway in vertebrates, but cannot explain the evolution of the strategy used by spiny lobsters or *Manduca.*

The Null Hypothesis

Perhaps the presence of a two-step transduction pathway in different taxa is coincidental, and any similarity is merely superficial. It may be misleading to focus on the observation that some olfactory receptor neurons contain ion channels that are gated by ions that enter during the transduction process, because a great diversity of transduction pathways has been described in olfactory receptor neurons [Dionne and Dubin, 1994; Schild and Restrepo, 1998; Ache and Restrepo, 2000]. We cannot at present dismiss the possibility that the presence of such a strategy in disparate groups simply reflects the fact that olfactory receptor neurons contain myriad ion channels that are directly or indirectly activated by odorants.

In summary, it is possible that a two-step transduction cascade is an adaptation for amplifying (or regulating the amplification of) small odorant signals, and has evolved independently in vertebrates, spiny lobsters, and *Manduca* for this purpose. However, we cannot conclude with any certainty that the use of two-step odorant transduction pathways constitutes an adaptation, because the function of this mechanism might differ among groups, and it seems possible that any similarity is coincidental.

Glomerular Neuropils

The axons of olfactory receptor cells terminate in glomeruli, large bundles of tangled neuropil, in the olfactory bulb of vertebrates, the olfactory lobe of lobsters, the antennal lobe of insects, and the tentacle ganglion in snails. These glomerular tangles usually have an overall round shape, and are encircled by glial cells.

Glomeruli in the vertebrate olfactory bulb consist of fibers emanating from peripheral receptor neurons, periglomerular interneurons, and various types of output cells. The cellular elements of glomerular circuits in some neopteran insects, such as *Manduca,* are organized in a remarkably similar fashion, comprising fibers from receptor neurons, interneurons, and antennal lobe output neurons [Hildebrand and Shepherd, 1997; Strausfeld and Hildebrand, 1999]. Many structural and functional features of glomeruli appear to be shared by vertebrates and insects, and I will describe these below. Nevertheless, it should be noted that most of the data available to date are drawn from studies of a small number of species: rats *(R. norvegicus),* mice *(M. musculus domesticus), Manduca,* and *Drosophila.* One feature that might be unique to insects is that chemosensory neurons from different portions of the body may send axons to glomerular structures in the antennal lobe; for example, in *Drosophila,* the axons of olfactory receptor neurons on the antennae and the maxillary palps both project to antennal lobe glomeruli, as do axons of olfactory receptor neurons on the antennae and in the labial pit organ of lepidopterans [Singh and Nayak, 1985; Kent et al., 1986].

In visual, auditory, and somatosensory systems, stimulus coding often involves the use of topographic maps in the central nervous system. This does not seem to be true of olfactory systems in any group examined to date: in both vertebrates and neopteran insects, there is no obvious relationship between the location of a glomerulus and the odorant stimuli to which it responds. Each glomerulus responds to a particular subset of odorants, such that a given odorant at a given concentration will evoke activity in the same subset of glomeruli in different individuals within a species, as has recently been demonstrated in imaging studies using rats, mice, zebrafish *(B. rerio),* honeybees *(Apis mellifera),* and two species of noctuid moths *(Heliothis virescens* and *H. zea)* [Friedrich and Korsching, 1997; Vickers et al., 1998; Galizia et al., 1999; Rubin and Katz, 1999; Wachowiak and Cohen, 2001]. In addition, inputs from the receptor epithelium to the glomerular layer appear to be tightly controlled and reproducible among members of a species. For example, in *Drosophila* as well as in rats and mice, axons of receptor cells that express a particular receptor or subset of receptors converge at the same glomerulus, which is in the same location in different individuals within a species [Ressler et al., 1994; Vassar et al., 1994; Mombaerts et al., 1996; Vosshall et al., 2000; Schaefer et al., 2001]. As described above, receptors are expressed in a few large zones in the vertebrate olfactory epithelium and on *Drosophila* antennae. A given glomerulus receives input from receptor neurons scattered throughout one of these zones, but there is no fine-scale topographical organization of inputs from the sensory periphery to glomeruli [Stocker et al., 1983; Clancy et al., 1994; Schoenfeld et al., 1994; Vosshall, 2001]. Thus, although we do not completely understand the role of glomeruli in coding odorant information, they appear to play a similar role in insects and vertebrates.

Finally, even the cellular interactions involved in the development of glomeruli are similar in insects and vertebrates. Studies of developing *Manduca* demonstrate that ingrowing axons from olfactory receptor neurons interact with each other to form small protoglomeruli, which do not require interaction with potential synaptic targets, such as the output cells of the antennal lobe, to develop [Oland and Tolbert, 1998]. Nevertheless, unless the protoglomerulus is rapidly surrounded by glial cells, it will dissolve and the projections will become diffuse [Oland and Tolbert, 1988; Oland et al., 1988; Baumann et al., 1996]. In developing and adult mice, axons of olfactory receptor cells will form glomeruli and interact with neurons in ectopic locations, and the development of glomerular tangles has been thought to be an intrinsic property of olfactory receptor cell axons [Graziadei and Kaplan, 1980; Graziadei and Samanen, 1980; Graziadei and Monti Graziadei, 1986]. Recent studies of rat embryos demonstrate that axons of receptor neurons initially form protoglomeruli, then interact with glia to form glomerular boundaries, and later interact with dendrites of interneurons and output neurons, just as in *Manduca* [Valverde et al., 1992; Bailey et al., 1999; Treloar et al., 1999].

Glomeruli in decapod crustaceans share similarities with those of insects and vertebrates, but also possess some unique features. In decapods, the axons of olfactory receptor neurons interact with fibers from interneurons and projection neurons in unusual cone-shaped glomeruli located in the olfactory lobe, as has been demonstrated in the Australian yabby *(Cherax destructor),* red swamp crayfish *(Procambarus clarkii),* American lobster *(Homarus americanus),* and Caribbean spiny lobster *(P. argus)* [Sandeman and Luff, 1973; Mellon and Munger, 1990; Schmidt et al., 1992a; Helluy et al., 1995]. In lobsters, the glomeruli of the olfactory lobe comprise three distinct horizontal zones or compartments that are innervated by different interneurons [Schmidt et al., 1992b; Langworthy et al., 1997; Schmidt and Ache, 1997]. Intraglomerular compartments have also been described in rats and mice, and may correlate with dendritic fields of different classes of periglomerular interneurons [Treloar et al., 1996; Kosaka et al., 1998; Kasowski et al., 1999]; if so, the subdivision of glomeruli into different compartments in which receptor and output neurons interact with different groups of interneurons might constitute another feature of olfactory glomeruli that is shared across phyla. In addition to this anatomical feature, a physiological characteristic of glomeruli is also shared in vertebrates and crustaceans: presynaptic inhibition of sensory input to glomeruli occurs in both vertebrates (turtles, *Terapene carolina*) and lobsters *(P. argus),* but the underlying mechanisms differ in the two groups [Wachowiak and Cohen, 1999]. One unique feature of the olfactory system in lobsters and crayfish is the presence of small, round glomeruli in the accessory lobe and other secondary olfactory regions of the deutocerebrum [Sandeman and Luff, 1973; Blaustein et al., 1988; Helluy et al., 1993]. This feature appears to be an evolutionary innovation, as the accessory lobe is only present in a subset of malacostracan crustaceans [Sandeman et al., 1993].

Although phylogenetically widespread in olfactory targets in the central nervous system, glomeruli are not a universal feature. For example, although glomeruli are present in the olfactory bulbs of all craniates examined to date, including Pacific hagfish *(Eptatretus stouti)* [Wicht and Northcutt, 1992], they are not present in near outgroups to craniates, such as amphioxus or larval tunicates [Bone, 1960; Vorontsova et al., 1997].

Among molluscs, olfactory pathways have been described in detail only in gastropods (snails and slugs), and the phylogenetic distribution of glomeruli is not clear. Glomeruli may not be present in cephalopods: the olfactory lobe of *Nautilus* is layered, and that of *Octopus* is smaller and has no apparent large-scale organization [Young, 1965, 1971]. Although glomeruli are present in the first olfactory target in snails, not all olfactory axons terminate in these glomeruli. In the giant African snail *Achatina fulica*, the axons of some olfactory receptor neurons project past the glomerular neuropils of the digit to terminate in the body of the tentacle ganglion, or even in the cerebral ganglion [Chase and Tolloczko, 1993]. The functional significance of extra-glomerular olfactory projections is not understood, but primary olfactory projections that bypass the olfactory bulb have also been described in most classes of vertebrates [Eisthen, 1997].

The distribution of olfactory glomeruli among arthropods is variable, and has been examined in detail by Strausfeld and colleagues [Strausfeld et al., 1995, 1998; Strausfeld, 1998; Strausfeld and Hildebrand, 1999]. Glomeruli are broadly present in neopteran insects, such as *Drosophila* and *Manduca,* but are not present in those that have secondarily lost odorant-sensitive antennae, such as diving beetles *(Dytiscus marginalis)* [Strausfeld et al., 1998]. Glomeruli are also absent in members of the sister groups to neopterans, such as mayflies, dragonflies, and damselflies [Strausfeld, 1998; Strausfeld and Hildebrand, 1999]. Glomeruli are not present in antenno-recipient areas in basal hexapods, such as silverfish, firebrats, or bristletails [Strausfeld and Hildebrand, 1999]. Among crustaceans, the sister group to hexapods, the presence of glomeruli is variable. The presence of cone-shaped glomeruli in the olfactory lobe is a feature unique to decapod crustaceans. In contrast, the olfactory lobe of isopods contains large, round glomeruli like those typically seen in neopteran insects, and basal branchiopods such as *Triops* appear to completely lack a specialized antennal neuropil [Strausfeld, 1998; Strausfeld et al., 1998]. The other major groups of arthropods are myriapods and chelicerates, and

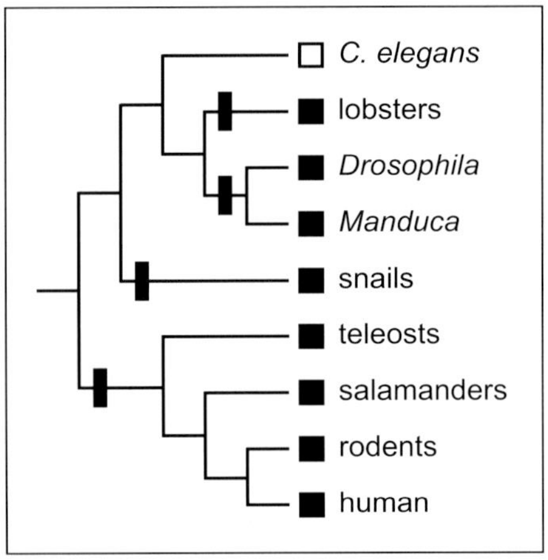

Fig. 4. Phylogenetic distribution of glomerular structures in the first central olfactory target in common model animals used in olfactory research. Taxa in which glomeruli have been described are indicated with solid boxes; the empty box signifies that *C. elegans* lacks glomeruli. Given that glomeruli are known to be lacking in outgroups relative to vertebrates, terrestrial insects, and lobsters, the hypothesis illustrated here postulates that glomerular structures evolved independently at least four times.

olfactory glomeruli are clearly present in at least some species in these groups. Among myriapods, glomeruli have been described in olfactory regions in chilopods (the centipede *Lithobius variegatus*) and diplopods (the desert millipede *Orthoporus ornatus*) [Strausfeld et al., 1995; Strausfeld, 1998]. Among chelicerates, olfactory glomeruli have been identified in various groups, including pycnogonids (sea spiders, *Lecythorhynchus hilgendorfii*), solpugids (sun spiders, *Eremobates pallipes*), scorpions (the bark scorpion *Centruroides sculpturatus*), opilionids (the group of false spiders containing the familiar 'daddy long-legs'), uropigids (vinegaroons or whip scorpions, *Mastigoproctus giganteus*), and amblypygids (tailless whip scorpions, *Tarantula*) [Strausfeld et al., 1998]. Glomeruli are not universally present within this group, and are lacking in the basal spider *Heptathela kimurai* [Strausfeld et al., 1998]. Interestingly, chelicerates have repeatedly coopted various legs for use as olfactory organs, and rather than being located in the most anterior segment, glomeruli tend to be present in the segmental ganglion that receives input from olfactory sensillae [Strausfeld et al., 1998]. Finally, olfactory glomeruli have been described in onychophorans (velvet worms), the sister group to arthropods: in *Euperipatoides leukartii,* afferents from antennal chemosensory neurons terminate in glomerular structures lateral to the mushroom bodies [Schürmann, 1995; Strausfeld et al., 1995].

The olfactory neurons of the amphid organ in *C. elegans* do not form glomerular structures, and seem to participate in circuits that are quite different from those in other phyla. Each olfactory neuron forms electrical and/or chemical synapses with interneurons and sometimes with other chemosensitive neurons, including, in some cases, its paired equivalent from the contralateral side [White et al., 1986]. The phylogenetic distribution of glomerular neuropils in olfactory regions of the central nervous system is illustrated in figure 4.

The similarity of architecture in the central olfactory targets of various animals is compelling, but a few notes of caution are in order. First, not all olfactory receptor neurons project to glomerular structures, and one must avoid circularity in identifying the olfactory component of chemosensory systems. For example, an insect may have chemosensitive sensillae on its legs, wings, mouth parts, and antennae. Which are 'olfactory'? Traditionally, the sensillae that respond to chemicals emanating from distant sources would be labeled olfactory, but making this functional determination can be difficult, and it is tempting to assume that the sensillae containing neurons with axons that project to glomerular structures constitute the olfactory system. This assumption can be misleading, as illustrated by studies of spiny lobsters *(P. argus).* Spiny lobsters have chemosensitive sensillae over most of their bodies, including the medial and lateral antennules, the antennae, and walking legs [Laverack, 1988]. The axons of neurons innervating aesthetasc sensillae on the medial antennule project to the 'olfactory lobe', which contains glomeruli, and those innervating non-aesthetasc sensillae on the medial and lateral antennules project to the 'lateral antennular neuropil', which lacks glomeruli [Schmidt et al., 1992a; Schmidt and Ache, 1996]. Despite the suggestive names of their central targets, both types of sensillae can mediate olfactory tasks, such as searching for and localizing food [Steullet et al., 2001].

In addition to the definition of 'olfaction', the definition of a 'glomerulus' can be problematic. Although the term was originally applied to structures in the olfactory bulb [Cajal, 1890], it has been widely used to refer to tangles of fibers [Pinching and Powell, 1971], or, even less specifically, to any 'synaptic complex enclosed in glial membranes or otherwise set apart' [Shepherd, 1974, p. 191]. By these definitions, many regions of the central nervous system might be interpreted as containing glomeruli [Leise, 1990]. Nevertheless, olfactory glomeruli possess unique features. Olfactory glomeruli are strikingly large, with diameters of 50–120 μm in rats *(R. norvegicus),* 45–100 μm in sphinx moths *(M. sexta),* and roughly 40–100 μm in snails *(Acha-*

tina fulica); in spiny lobsters *(P. argus),* the cone-shaped olfactory lobe glomeruli are 40–100 μm in diameter and 250 μm long [Pinching and Powell, 1971; Chase, 1985; Schmidt et al., 1992b; Rospars and Hildebrand, 2000]. Another key difference between glomeruli in olfactory centers and glomerulus-like structures in other portions of the brain, such as the large 'barrels' in the whisker region of rat somatosensory cortex, is that olfactory glomeruli do not receive topographically-organized afferent input, but instead seem to receive input from groups of axons expressing identical receptor genes. Thus, olfactory glomeruli possess features that distinguish them from other types of compartments in the central nervous system.

Overall, glomeruli are broadly present in olfactory targets in the central nervous system. Why? Is the presence of these odd structures significant for our understanding of olfactory function?

Homology

Given their widespread phylogenetic distribution glomeruli could have evolved once in the common ancestor of arthropods, molluscs, and craniates. However, given that glomeruli are not present in outgroups relative to craniates or neopteran insects, and are absent in basal crustaceans, glomeruli must have arisen independently at least three times, and cannot be considered homologous across groups. Further, the phylogenetic distribution of glomeruli among arthropods suggests that glomeruli have evolved independently several times within this group [Strausfeld et al., 1995, 1998; Strausfeld, 1998; Strausfeld and Hildebrand, 1999]. We can therefore reject the hypothesis that glomeruli are homologous across phyla.

Convergence

The repeated evolution of olfactory glomeruli might be due to evolutionary convergence. If so, are glomeruli a functional adaptation for processing odorant information, or a response to a constraint?

Adaptation. If glomeruli arose many times independently as an adaptation related to olfactory information processing, their presence in diverse taxa is an important clue to understanding the neurobiology of olfaction. The repeated evolution of these enigmatic structures has fascinated olfactory researchers, most of whom accept the hypothesis that glomeruli play a crucial role in the coding of odorant information [Hildebrand and Shepherd, 1997; Strausfeld and Hildebrand, 1999; Christensen and White, 2000]. Unlike visual or auditory stimuli, olfactory stimuli cannot be arrayed along a small number of linear dimensions; unlike the visual or auditory systems, the olfactory system does not use topographic maps to represent the stimulus, but may instead use glomerular structures for a similar purpose. The exact function of glomeruli, and their role in odorant information coding, remains unclear, but two fundamental roles for glomeruli have been proposed. First, the architecture of glomeruli might function to amplify odorant signals, as has been demonstrated in studies with both insects and vertebrates [Duchamp-Viret et al., 1989, 1990; Hartlieb et al., 1997]. Another hypothesis holds that glomeruli and their associated interneurons may provide for a form of lateral inhibition, or could serve to sharpen the contrast among different odorants [Shepherd, 1974, 1992]. These hypotheses are not mutually exclusive, and both mechanisms could work together to enhance the signal-to-noise ratio between the olfactory periphery and the central nervous system.

Constraint. The presence of glomeruli in olfactory targets may reflect the operation of physical constraints imposed by the need to have similarly-tuned receptor neurons scattered throughout regions of the olfactory epithelium, either to protect against complete loss or to ensure that the odorant signal is averaged across as large an epithelial sheet as possible. The axons of these scattered neurons then must converge on a small number of locations in the central nervous system. In this view, glomerular structures could simply be a space-efficient method of bringing together axons of similarly-tuned receptor neurons and segregating them from other aggregations of axons. Hildebrand and Shepherd [1997] argue persuasively that if this were the case, interneurons between glomeruli would be unnecessary; nevertheless, such neurons are found in vertebrates, insects, and lobsters.

The Null Hypothesis

Although it is conceivable that the similar organization of central olfactory targets into glomerular structures in different taxa is coincidental, or that the similarities are superficial, the many shared morphological and physiological features of glomeruli make this hypothesis seem unlikely. It is also possible that glomeruli serve different functions in different groups; again, the many similarities make this seem unlikely, but until we develop a better understanding of the function of glomeruli, we cannot completely dismiss this hypothesis.

In summary, glomeruli have arisen repeatedly many times in the olfactory systems of a wide array of animals. Data currently available suggest that glomeruli constitute a functional adaptation for processing odorant information, although their precise role in olfactory information coding remains unclear.

Implications for Olfactory Research

Researchers studying the neurobiology of olfaction commonly assume that features that are present in distantly-related animals are functional adaptations for carrying or processing odorant information. The analysis presented here demonstrates that other hypotheses can rarely be ruled out and that in many cases convergent features could have been shaped by constraints as well as adaptation. The preceding analysis suggests that a consideration of alternative hypotheses based on an explicit evolutionary framework is a useful method for determining which hypotheses logically compete with each other and for developing new hypotheses.

For example, odorant binding proteins (OBPs) may be present in terrestrial animals because such animals face the constraint of having to detect hydrophobic odorants in air; conversely, aquatic animals may be constrained from using OBPs because of the high metabolic cost of producing soluble proteins that could diffuse away into the environment. In addition, OBPs might serve one or more functional roles in carrying odorants to the receptor, interacting with odorant receptors, or removing odorants from the sensory epithelium. Because the function of OBPs is not yet known, we cannot dismiss the hypothesis that the presence of these proteins in insects and mammals is coincidental, or that OBPs serve a role not specific to olfaction in one or both groups.

The presence of a two-step signal transduction strategy in vertebrates, spiny lobsters, and sphinx moths *(Manduca)* could represent an adaptation for detecting small quantities of odorants. However, given the great diversity of transduction mechanisms employed in olfactory receptor neurons, the similarity may be coincidental, and these mechanisms might also have arisen for different reasons in different taxa.

The clearest examples of adaptive convergence in olfactory systems appear to be the use of G protein-coupled receptors (GPCRs) as odorant receptors in *C. elegans, Drosophila,* and vertebrates, and the presence of glomerular structures in the first olfactory relay in the central nervous systems of many animals. GPCRs and their underlying genes seem to be remarkably flexible, which might explain why they been co-opted repeatedly for use in olfaction. The repeated evolutionary origin of olfactory glomeruli strongly suggests that this anatomical arrangement serves a valuable and specific function in olfactory information processing, although an understanding of the precise function of these structures remains elusive.

Implications for the Use of Convergence in Neurobiological Research

Many neurobiologists use a 'model organism' approach in which similarity is emphasized without consideration of the evolutionary processes that may have given rise to these similarities. Similar features can be present due to inheritance from a common ancestor or may represent examples of convergent evolution. In the latter case, these features may represent responses to similar constraints, or could constitute adaptations to similar demands. The interpretation of similar features depends on an explicit or implicit hypothesis about the underlying cause of similarity, and an explicit attention to evolutionary processes would enrich our understanding of many problems in neurobiology.

The goal of this symposium was to draw attention to examples of convergent evolution, which can serve as valuable clues to understanding strategies for coding information in the nervous system. As demonstrated in this paper, adaptation and constraints may operate together to shape many features, and unless we know the function of a feature we often cannot exclude the possibility that any similarity is superficial or misleading. Thus, we should not simply assume that features that have arisen independently constitute adaptations that will prove informative about mechanisms of neural processing.

Acknowledgments

In developing the ideas presented here, I have benefited enormously from conversations with Barry Ache, Curt Bell, Vince Dionne, Rona Delay, Chuck Derby, Mike Grotewiel, Jack Johnson, Barb Lundrigan, Jeanette McGuire, Daesik Park, Kurt Schwenk, Dick Vogt, and Sarah Zawacki. I thank them all for their generosity in answering questions and providing ideas and suggestions. I am particularly grateful to Nick Strausfeld for many enlightening and entertaining discussions of convergence in olfactory systems, and to Kiisa Nishikawa and Daesik Park for comments on the manuscript. Of course, none of these people is to blame for anything I say in this paper. Finally, I thank Dr. Thomas Karger and S. Karger Publishing for their continuing support of the annual Karger Workshops and the J.B. Johnston Club.

References

Ache, B.W. (1994) Towards a common strategy for transducing olfactory information. Sem. Cell Biol., *5:* 55–63.

Ache, B.W., and D. Restrepo (2000) Olfactory transduction. *In* The Neurobiology of Taste and Smell (ed. by T.E. Finger, W.L. Silver and D. Restrepo), Wiley-Liss, New York, pp. 159–177.

Anderson, P.A., and B.W. Ache (1985) Voltage- and current-clamp recordings of the receptor potential in olfactory receptor cells *in situ.* Brain Res., *338*: 273–280.

Ayer, R.K., Jr., and J. Carlson (1992) Olfactory physiology in the *Drosophila* antenna and maxillary palp: *acj6* distinguishes two classes of odorant pathways. J. Neurobiol., *23:* 965–982.

Bacchini, A., E. Gaetani, and A. Cavaggioni (1992) Pheromone binding proteins of the mouse, *Mus musculus.* Experientia, *48:* 419–421.

Bailey, M.S., A.C. Puche, and M.T. Shipley (1999) Development of the olfactory bulb: Evidence for glia-neuron interactions in glomerular formation. J. Comp. Neurol., *415:* 423–448.

Bakalyar, H.A., and R.R. Reed (1990) Identification of a specialized adenylyl cyclase that may mediate odorant detection. Science, *250:* 1403–1406.

Baldaccini, N.E., A. Gagliardo, P. Pelosi, and A. Topazzini (1986) Occurrence of a pyrazine binding protein in the nasal mucosa of some vertebrates. Comp. Biochem. Physiol. B, *84:* 249–253.

Bargmann, C.I. (1998) Neurobiology of the *Caenorhabditis elegans* genome. Science, *282:* 2028–2033.

Bargmann, C.I., E. Hartwieg, and H.R. Horvitz (1993) Odorant-selective genes and neurons mediate olfaction in C. elegans. Cell, *13:* 515–527.

Barth, A.L., N.J. Justice, and J. Ngai (1996) Asynchronous onset of odorant receptor expression in the developing zebrafish olfactory system. Neuron, *16:* 23–34.

Baumann, P.M., L.A. Oland, and L.P. Tolbert (1996) Glial cells stabilize axonal protoglomeruli in the developing olfactory lobe of the moth *Manduca sexta.* J. Comp. Neurol., *373:* 118–128.

Bell, C.C. (2002) Evolution of cerebellum-like structures. Brain Behav. Evol., *59:* 312–326.

Belluscio, L., G.H. Gold, A. Nemes, and R. Axel (1998) Mice deficient in G_{olf} are anosmic. Neuron, *20:* 69–81.

Berger, L. (1967) Embrional and larval development of F1 generation of green frogs different combinations. Acta Zool. Cracov., *12:* 123–160.

Berger, L. (1968) Morphology of the F1 generation of various crosses within *Rana esculenta* complex. Acta Zool. Cracov., *13:* 301–324.

Bignetti, E., G. Damiani, P. De Negri, R. Ramoni, F. Avanzini, G. Ferrari, and G.L. Rossi (1987) Specificity of an immunoaffinity column for odorant-binding protein from bovine nasal mucosa. Chem. Senses, *12:* 601–608.

Blaustein, D.N., C.D. Derby, R.B. Simmons, and A.C. Beall (1988) Structure of the brain and medulla terminalis of the spiny lobster *Panulirus argus* and the crayfish *Procambrus clarkii* with an emphasis on olfactory centers. J. Crust. Biol., *8:* 493–519.

Böcskei, Z., C.R. Groom, D.R. Flower, C.E. Wright, S.E.V. Phillips, A. Cavaggioni, J.B.C. Findlay, and A.C.T. North (1992) Pheromone binding to two rodent urinary proteins revealed by X-ray crystallography. Nature, *360:* 186–188.

Boekhoff, I., W.C. Michel, H. Breer, and B.W. Ache (1994) Single odors differentially stimulate dual second messenger pathways in lobster olfactory receptor cells. J. Neurosci., *14:* 3304–3309.

Boekhoff, I., K. Raming, and H. Breer (1990a) Pheromone-induced stimulation of inositol-trisphosphate formation in insect antennae is mediated by G-proteins. J. Comp. Physiol. B, *160:* 99–103.

Boekhoff, I., J. Strotmann, K. Raming, E. Tareilus, and H. Breer (1990b) Odorant-sensitive phospholipase C in insect antennae. Cell Signal, *2:* 49–56.

Bone, Q. (1960) The central nervous system in amphioxus. J. Comp. Neurol., *115:* 27–64.

Boudjelal, M., A. Sivaprasadarao, and J.B.C. Findlay (1996) Membrane receptor for odour-binding proteins. Biochem. J., *317:* 23–27.

Bouvier, M. (2001) Oligomerization of G-protein-coupled transmitter receptors. Nat. Rev. Neurosci., *2:* 274–286.

Breer, H., I. Boekhoff, and E. Tareilus (1990) Rapid kinetics of second messenger formation in olfactory transduction. Nature, *345:* 65–68.

Brunet, L.J., G.H. Gold, and J. Ngai (1996) General anosmia caused by a targeted disruption of the mouse olfactory cyclic nucleotide-gated cation channel. Neuron, *17:* 681–693.

Buck, L., and R. Axel (1991) A novel multigene family may encode odorant receptors: A molecular basis for odor recognition. Cell, *65:* 175–187.

Cajal, S.R. (1890) Origen y terminación de las fibras nerviosas olfatorias. Gazz. Sanit. Barcelona, 1–21.

Chase, R. (1985) Responses to odors mapped in snail tentacle and brain by [^{14}C]-2-deoxyglucose autoradiography. J. Neurosci., *5:* 2930–2939.

Chase, R., and B. Tolloczko (1993) Tracing neural pathways in snail olfaction: From the tip of the tentacles to the brain and beyond. Microsc. Res. Tech., *24:* 214–230.

Chess, A., I. Simon, H. Cedar, and R. Axel (1994) Allelic inactivation regulates olfactory receptor gene expression. Cell, *78:* 823–834.

Christensen, T.A., and J. White (2000) Representation of olfactory information in the brain. *In* The Neurobiology of Taste and Smell (ed. by T.E. Finger, W.L. Silver and D. Restrepo), Wiley-Liss, New York, pp. 201–232.

Clancy, A.N., T.A. Schoenfeld, W.B. Forbes, and F. Macrides (1994) The spatial organization of the peripheral olfactory system of the hamster. Part II: Receptor surfaces and odorant passageways within the nasal cavity. Brain Res. Bull., *34:* 211–241.

Clyne, P.J., C.G. Warr, M.R. Freeman, D. Lessing, J. Kim, and J.R. Carlson (1999) A novel family of divergent seven-transmembrane proteins: Candidate odorant receptors in *Drosophila.* Neuron, *22:* 203–204.

Coburn, C.M., and C.I. Bargmann (1996) A putative cyclic nucleotide-gated channel is required for sensory development and function in *C. elegans.* Neuron, *17:* 695–706.

Dhallen, R.S., K.-W. Yau, K.A. Schrader, and R.R. Reed (1990) Primary structure and functional expression of a cyclic nucleotide-activated channel from olfactory neurons. Nature, *347:* 184–187.

Dionne, V.E., and A.E. Dubin (1994) Transduction diversity in olfaction. J. Exp. Biol., *194:* 1–21.

Doolittle, R.F. (1994) Convergent evolution: The need to be explicit. Trends Biochem. Sci., *19:* 15–18.

Du, G., and G.D. Prestwich (1995) Protein structure encodes the ligand binding specificity in pheromone binding proteins. Biochemistry, *34:* 8726–8732.

Dubin, A.E., and V.E. Dionne (1993) Modulation of Cl^-, K^+, and nonselective cation conductances by taurine in olfactory receptor neurons of the mudpuppy *Necturus maculosus.* J. Gen. Physiol., *101:* 469–485.

Duchamp-Viret, P., A. Duchamp, and G. Sicard (1990) Olfactory discrimination over a wide concentration range. Comparison of receptor cell and bulb neuron abilities. Brain Res., *517:* 256–262.

Duchamp-Viret, P., A. Duchamp, and M. Vigouroux (1989) Amplifying role of convergence in olfactory system. A comparative study of receptor cell and second-order neuron sensitivities. J. Neurophysiol., *61:* 1085–1094.

Eisthen, H.L. (1997) Evolution of vertebrate olfactory systems. Brain Behav. Evol., *50:* 222–233.

Eldredge, N., and J. Cracraft (1980) Phylogenetic Patterns and the Evolutionary Process: Method and Theory in Comparative Biology. Columbia University Press, New York.

Fadool, D.A., and B.W. Ache (1992) Plasma membrane inositol 1,4,5-trisphosphate-activated channels mediate signal transduction in lobster olfactory receptor neurons. Neuron, *9:* 907–918.

Fadool, D.A., S.J. Estey, and B.W. Ache (1995) Evidence that a G_q-protein mediates excitatory odor transduction in lobster olfactory receptor neurons. Chem. Senses, *20:* 489–498.

Felicioli, A., M. Ganni, M. Garibotti, and P. Pelosi (1993) Multiple types and forms of odorant-binding proteins in the Old-World porcupine *Hystrix cristata.* Comp. Biochem. Physiol. B, *105:* 775–784.

Finlay, B.L., and R.B. Darlington (1995) Linked regularities in the development and evolution of mammalian brains. Science, *268:* 1578–1584.

Firestein, S., B. Darrow, and G.M. Shepherd (1991) Activation of the sensory current in salamander olfactory receptor neurons depends on a G protein-mediated cAMP second-messenger system. Neuron, *6:* 825–835.

Freitag, J., A. Beck, G. Ludwig, L. von Buchholtz, and H. Breer (1999) On the origin of the olfactory receptor family: Receptor genes of the jawless fish *(Lampetra fluviatilis)*. Gene, *226:* 165–174.

Freitag, J., J. Krieger, J. Strotmann, and H. Breer (1995) Two classes of olfactory receptors in *Xenopus laevis*. Neuron, *15:* 1383–1392.

Freitag, J., G. Ludwig, I. Andreini, P. Rössler, and H. Breer (1998) Olfactory receptors in aquatic and terrestrial vertebrates. J. Comp. Physiol. A, *183:* 635–650.

Friedrich, R.W., and S.I. Korsching (1997) Combinatorial and chemotopic odorant coding in the zebrafish olfactory bulb visualized by optical imaging. Neuron, *18:* 737–752.

Frings, S., and B. Lindemann (1991) Current recording from sensory cilia of olfactory receptor cells in situ. 1. The neuronal response to cyclic nucleotides. J. Gen. Physiol., *97:* 1–16.

Galizia, C.G., S. Sachse, A. Rappert, and R. Menzel (1999) The glomerular code for odor representation is species specific in the honeybee *Apis mellifera*. Nat. Neurosci., *2:* 473–478.

Gao, Q., and A. Chess (1999) Identification of candidate *Drosophila* olfactory receptors from genomic DNA sequence. Genomics, *60:* 31–39.

Garibotti, M., A. Navarrini, A.M. Pisanelli, and P. Pelosi (1997) Three odorant-binding proteins from rabbit nasal mucosa. Chem. Senses, *22:* 383–390.

Gomez, G., N.E. Rawson, C.G. Hahn, R. Michaels, and D. Restrepo (2000) Characteristics of odorant elicited calcium changes in cultured human olfactory neurons. J. Neurosci. Res., *62:* 737–749.

Gould, S.J., and E.S. Vrba (1982) Exaptation – A missing term in the science of form. Paleobiology, *8:* 4–15.

Goulding, E.H., J. Ngai, R.H. Kramer, S. Colicos, R. Axel, S.A. Siegelbaum, and A. Chess (1992) Molecular cloning and single-channel properties of the cyclic nucleotide-gated channel from catfish olfactory neurons. Neuron, *8:* 45–58.

Graziadei, P.P.C., and M.S. Kaplan (1980) Regrowth of olfactory sensory axons into transplanted neural tissue. 1. Development of connections with the occipital cortex. Brain Res., *201:* 39–44.

Graziadei, P.P.C., and G.A. Monti Graziadei (1986) Principles of organization of the vertebrate olfactory glomerulus: An hypothesis. Neuroscience, *19:* 1025–1035.

Graziadei, P.P.C., and D.W. Samanen (1980) Ectopic glomerular structures in the olfactory bulb of neonatal and adult mice. Brain Res., *187:* 467–472.

Hartlieb, E., S. Anton, and B.S. Hansson (1997) Dose-dependent response characteristics of antennal lobe neurons in the male moth *Agrotis segetum* (Lepidoptera: Noctuidae). J. Comp. Physiol. A, *181:* 469–476.

Hatt, H., and B.W. Ache (1994) Cyclic nucleotide- and inositol phosphate-gated ion channels in lobster olfactory receptor neurons. Proc. Natl. Acad. Sci. USA, *91:* 6264–6268.

Hekmat-Scafe, D.S., R.A. Steinbrecht, and J.R. Carlson (1997) Coexpression of two odorant-binding protein homologs in *Drosophila:* Implications for olfactory coding. J. Neurosci., *17:* 1616–1624.

Helluy, S., R. Sandeman, B. Beltz, and D. Sandeman (1993) Comparative brain ontogeny of the crayfish and clawed lobster: Implications of direct and larval development. J. Comp. Neurol., *335:* 343–354.

Helluy, S.M., M.L. Ruchhoeft, and B.S. Beltz (1995) Development of the olfactory and accessory lobes in the American lobster: An allometric analysis and its implications for the deutocerebral structure of decapods. J. Comp. Neurol., *357:* 433–445.

Hildebrand, J.G., and G.M. Shepherd (1997) Mechanisms of olfactory discrimination: Converging evidence for common principles across phyla. Ann. Rev. Neurosci., *20:* 595–631.

Jansen, G., K.L. Thijssen, P. Werner, M. van der Horst, E. Hazendonk, and R.H. Plasterk (1999) The complete family of genes encoding G proteins of *Caenorhabditis elegans*. Nat. Genet., *21:* 414–419.

Jones, D.T., and R.R. Reed (1989) G_{olf}: An olfactory neuron specific-G protein involved in odorant signal transduction. Science, *244:* 790–795.

Josefsson, L.G. (1999) Evidence for kinship between diverse G-protein coupled receptors. Gene, *239:* 333–340.

Kaissling, K.E. (1986) Chemo-electrical transduction in insect olfactory receptors. Ann. Rev. Neurosci., *9:* 121–145.

Kaissling, K.E. (1998) Pheromone deactivation catalyzed by receptor molecules: A quantitative kinetic model. Chem. Senses, *23:* 385–395.

Kasowski, H.J., H. Kim, and C.A. Greer (1999) Compartmental organization of the olfactory bulb glomerulus. J. Comp. Neurol., *407:* 261–274.

Kent, K.S., I.D. Harrow, P. Quartararo, and J.G. Hildebrand (1986) An accessory olfactory pathway in Lepidoptera: The labial pit organ and its central projections in *Manduca sexta* and certain other sphinx moths and silk moths. Cell Tissue Res., *245:* 237–245.

Kleene, S.J. (1993) Origin of the chloride current in olfactory transduction. Neuron, *11:* 123–132.

Kleene, S.J., and R.C. Gesteland (1991) Calcium-activated chloride conductance in frog olfactory cilia. J. Neurosci., *11:* 3624–3629.

Kolesnikov, S.S., and A.V. Kosolapov (1993) Cyclic nucleotide-activated channels in carp olfactory receptor cells. Biochim. Biophys. Acta, *1150:* 63–72.

Komatsu, H., Y.H. Jin, N. L'Etoile, I. Mori, C.I. Bargmann, N. Akaike, and Y. Ohshima (1999) Functional reconstitution of a heteromeric cyclic nucleotide-gated channel of *Caenorhabditis elegans* in cultured cells. Brain Res., *821:* 160–168.

Komatsu, H., I. Mori, J.S. Rhee, N. Akaike, and Y. Ohshima (1996) Mutations in a cyclic nucleotide-gated channel lead to abnormal thermosensation and chemosensation in C. elegans. Neuron, *17:* 707–718.

Kosaka, K., K. Toida, Y. Aika, and T. Kosaka (1998) How simple is the organization of the olfactory glomerulus?: The heterogeneity of so-called periglomerular cells. Neurosci. Res., *30:* 101–110.

Krautwurst, D., K.-W. Yau, and R.R. Reed (1998) Identification of ligands for olfactory receptors by functional expression of a receptor library. Cell, *95:* 917–926.

Kurahashi, T., and K.-W. Yau (1993) Co-existence of cationic and chloride components in odorant-induced current of vertebrate olfactory receptor cells. Nature, *363:* 71–74.

Lacazette, E., A.M. Gachon, and G. Pitiot (2000) A novel human odorant-binding protein gene family resulting from genomic duplicons at 9q34:Differential expression in the oral and genital spheres. Hum. Mol. Genet., *22:* 289–301.

Langworthy, K., S. Helluy, J. Benton, and B. Beltz (1997) Amines and peptides in the brain of the American lobster: Immunocytochemical localization patterns and implications for brain function. Cell Tissue Res., *288:* 191–206.

Laverack, M.S. (1988) The diversity of chemoreceptors. *In* Sensory Biology of Aquatic Animals (ed. by J. Atema, R.R. Fay, A.N. Popper and W.N. Tavolga), Springer, New York, pp. 287–312.

Lee, K.H., R.G. Wells, and R.R. Reed (1987) Isolation of an olfactory cDNA: Similarity to retinol-binding protein suggests a role in olfaction. Science, *235:* 1053–1056.

Leise, E.M. (1990) Modular construction of nervous systems: A basic principle of design for invertebrates and vertebrates. Brain Res. Rev., *15:* 1–23.

Levy, N.S., H.A. Bakalyar, and R.R. Reed (1991) Signal transduction in olfactory neurons. J. Steroid Biochem. Mol. Biol., *39:* 633–637.

Löbel, D., M. Jacob, M. Volkner, and H. Breer (2002) Odorants of different chemical classes interact with distinct odorant binding protein subtypes. Chem. Senses, *27:* 39–44.

Löbel, D., J. Strotmann, M. Jacob, and H. Breer (2001) Identification of a third rat odorant-binding protein (OBP3). Chem. Senses, *26:* 673–680.

Lowe, G., and G.H. Gold (1993) Nonlinear amplification by calcium-dependent chloride channels in olfactory receptor cells. Nature, *366:* 283–286.

Ludwig, J., T. Margalit, E. Eismann, D. Lancet, and U.B. Kaupp (1990) Primary structure of cAMP-gated channel from bovine olfactory epithelium. FEBS Lett., *270:* 24–29.

Ma, L., and W.C. Michel (1998) Drugs affecting phospholipase C-mediated signal transduction block the olfactory cyclic nucleotide-gated current of adult zebrafish. J. Neurophysiol., *79:* 1183–1192.

Max, M., Y.G. Shanker, L. Huang, M. Rong, Z. Liu, F. Campagne, H. Weinstein, S. Damak, and R.F. Margolskee (2001) *Tas1r3,* encoding a new candidate taste receptor, is allelic to the sweet responsiveness locus *Sac.* Nat. Genet., *28:* 58–63.

Maynard Smith, J., R. Burian, S. Kauffman, P. Alberch, J. Campbell, B. Goodwin, R. Lande, D. Raup, and L. Wolpert (1985) Developmental constraints and evolution. Q. Rev. Biol., *60:* 265–287.

Mellon, D.J., and S.D. Munger (1990) Nontopographic projection of olfactory sensory neurons in the crayfish brain. J. Comp. Neurol., *296:* 253–262.

Menini, A., C. Picco, and S. Firestein (1995) Quantal-like current fluctuations induced by odorants in olfactory receptor cells. Nature, *373:* 435–437.

Mombaerts, P., F. Wang, C. Dulac, S.K. Chao, A. Nemes, M. Mendelsohn, J. Edmondson, and R. Axel (1996) Visualizing an olfactory sensory map. Cell, *87:* 675–686.

Mori, I., and Y. Ohshima (1997) Molecular neurogenetics of chemotaxis and thermotaxis in the nematode *Caenorhabditis elegans.* Bioessays, *19:* 1055–1064.

Mucignat-Caretta, C., A. Caretta, and A. Cavaggioni (1995) Acceleration of puberty onset in female mice by male urinary proteins. J. Physiol., *486:* 517–522.

Munger, S.D., R.A. Gleeson, H.C. Aldrich, N.C. Rust, B.W. Ache, and R.M. Greenberg (2000) Characterization of a phosphoinositide-mediated odor transduction pathway reveals plasma membrane localization of an inositol 1,4, 5-trisphosphate receptor in lobster olfactory receptor neurons. J. Biol. Chem., *275:* 20450–20457.

Murrell, J.R., and D.D. Hunter (1999) An olfactory sensory neuron line, *odora,* properly targets olfactory proteins and responds to odorants. J. Neurosci., *19:* 8260–8270.

Nakamura, T., and G.H. Gold (1987) A cyclic nucleotide-gated conductance in olfactory receptor cilia. Nature, *325:* 442–444.

Nef, P., I. Hermansborg-Meyer, H. Artières-Pin, L. Beasley, V.E. Dionne, and S.F. Heinemann (1992) Spatial pattern of receptor expression in the olfactory epithelium. Proc. Natl. Acad. Sci. USA, *89:* 8948–8952.

Nef, S., I. Allaman, H. Fiumelli, E. De Castro, and P. Nef (1996) Olfaction in birds: Differential embryonic expression of nine putative odorant receptor genes in the avian olfactory system. Mech. Dev., *55:* 65–77.

Nelson, G., M.A. Hoon, J. Chandrashekar, Y. Zhang, N.J. Ryba, and C.S. Zuker (2001) Mammalian sweet taste receptors. Cell, *106:* 381–390.

Ngai, J., M.M. Dowling, L. Buck, R. Axel, and A. Chess (1993a) The family of genes encoding odorant receptors in the channel catfish. Cell, *72:* 657–666.

Ngai, J., A. Chess, M.M. Dowling, N. Necles, E.R. Macagno, and R. Axel (1993b) Coding of olfactory information: Topography of odorant receptor expression in the catfish olfactory epithelium. Cell, *72:* 667–680.

Nishikawa, K.C. (2002) Evolutionary convergence in nervous systems: Insights from comparative phylogenetic studies. Brain Behav. Evol., *59:* 240–249.

Oland, L.A., and L.P. Tolbert (1988) Effects of hydroxyurea parallel the effects of radiation in developing olfactory glomeruli in insects. J. Comp. Neurol., *278:* 377–387.

Oland, L.A., and L.P. Tolbert (1998) Glomerulus development in the absence of a set of mitral-like neurons in the insect olfactory lobe. J. Neurobiol., *36:* 41–52.

Oland, L.A., L.P. Tolbert, and K.L. Mossman (1988) Radiation-induced reduction of the glial population during development disrupts the formation of olfactory glomeruli in an insect. J. Neurosci., *8:* 353–367.

Pelosi, P., N.E. Baldaccini, and A.M. Pisanelli (1982) Identification of a specific olfactory receptor for 2-isobutyl-3-methoxypyrazine. Biochem. J., *201:* 245–248.

Pevsner, J., R.R. Trifiletti, S.M. Strittmatter, and S.H. Snyder (1985) Isolation and characterization of an olfactory receptor protein for odorant pyrazines. Proc. Natl. Acad. Sci. USA, *82:* 3050–3054.

Pinching, A.J., and T.P.S. Powell (1971) The neuropil of the glomeruli of the olfactory bulb. J. Cell. Sci., *9:* 347–377.

Plettner, E., J. Lazar, E.G. Prestwich, and G.D. Prestwich (2000) Discrimination of pheromone enantiomers by two pheromone binding proteins from the gypsy moth *Lymantria dispar.* Biochemistry, *39:* 8953–8962.

Prestwich, G.D., G. Du, and S. LaForest (1995) How is pheromone specificity encoded in proteins? Chem. Senses, *20:* 461–469.

Ressler, K.J., S.L. Sullivan, and L.B. Buck (1993) A zonal organization of odorant receptor gene expression in the olfactory epithelium. Cell, *73:* 597–609.

Ressler, K.J., S.L. Sullivan, and L.B. Buck (1994) Information coding in the olfactory system: Evidence for a stereotyped and highly organized epitope map in the olfactory bulb. Cell, *79:* 1245–1255.

Reuter, D., K. Zierold, W.H. Schröder, and S. Frings (1998) A depolarizing chloride current contributes to chemoelectrical transduction in olfactory sensory neurons *in situ.* J. Neurosci., *18:* 6623–6630.

Riesgo-Escovar, J., D. Raha, and J.R. Carlson (1995) Requirement for a phospholipase C in odor response: Overlap between olfaction and vision in *Drosophila.* Proc. Natl. Acad. Sci. USA, *92:* 2864–2868.

Ringo, J.L. (1991) Neuronal connections as a function of brain size. Brain Behav. Evol., *38:* 1–6.

Robertson, D.H.L., R.J. Beynon, and R.P. Evershed (1993) Extraction, characterization, and binding analysis of two pheromonally active ligands associated with major urinary protein of house mouse *(Mus musculus).* J. Chem. Ecol., *19:* 1405–1416.

Robertson, H.M., R. Martos, C.R. Sears, E.Z. Todres, K.K. Walden, and J.B. Nardi (1999) Diversity of odourant binding proteins revealed by an expressed sequence tag project on male *Manduca sexta* moth antennae. Insect Mol. Biol., *8:* 501–518.

Rogers, M.E., M. Sun, M.R. Lerner, and R.G. Vogt (1997) Snmp-1, a novel membrane protein of olfactory neurons of the silk moth *Antheraea polyphemus* with homology to the CD36 family of membrane proteins. J. Biol. Chem., *272:* 14792–14799.

Rospars, J.P., and J.G. Hildebrand (2000) Sexually dimorphic and isomorphic glomeruli in the antennal lobes of the sphinx moth *Manduca sexta.* Chem. Senses, *25:* 119–129.

Rubin, B.D., and L.C. Katz (1999) Optical imaging of odorant representations in the mammalian olfactory bulb. Neuron, *23:* 499–511.

Sandeman, D.C., and S.E. Luff (1973) The structural organization of glomerular neuropile in the olfactory and accessory lobes of an Australian freshwater crayfish, *Cherax destructor.* Z. Zellforsch. Mikroskop. Anat., *142:* 37–61.

Sandeman, D.C., G. Scholtz, and R.E. Sandeman (1993) Brain evolution in decapod crustacea. J. Exp. Zool., *265:* 112–133.

Sato, K., and N. Suzuki (2000) The contribution of a Ca^{2+}-activated Cl^- conductance to amino-acid-induced inward current responses of ciliated olfactory neurons of the rainbow trout. J. Exp. Biol., *203:* 253–262.

Schaefer, M.L., T.E. Finger, and D. Restrepo (2001) Variability of position of the P2 glomerulus within a map of the mouse olfactory bulb. J. Comp. Neurol., *436:* 351–362.

Schild, D., and D. Restrepo (1998) Transduction mechanisms in vertebrate olfactory receptor cells. Physiol. Rev., *78:* 429–466.

Schmidt, M., and B.W. Ache (1996) Processing of antennular input in the brain of the spiny lobster, *Panulirus argus.* I. Non-olfactory chemosensory and mechanosensory pathway of the lateral and median antennular neuropils. J. Comp. Physiol. A, *178:* 579–604.

Schmidt, M., and B.W. Ache (1997) Immunocytochemical analysis of glomerular regionalization and neuronal diversity in the olfactory deutocerebrum of the spiny lobster. Cell Tissue Res., *287:* 541–563.

Schmidt, M., L. van Ekeris, and B. Ache (1992a) Antennular projections to the midbrain of the spiny lobster. I. Sensory innervation of the lateral and medial antennular neuropils. J. Comp. Neurol., *318:* 277–290.

Schmidt, M., L. van Ekeris, and B. Ache (1992b) Antennular projections to the midbrain of the spiny lobster. II. Sensory innervation of the olfactory lobe. J. Comp. Neurol., *318:* 291–303.

Schoenfeld, T.A., A.N. Clancy, W.B. Forbes, and F. Macrides (1994) The spatial organization of the peripheral olfactory system of the hamster. Part I: Receptor neuron projections to the main olfactory bulb. Brain Res. Bull., *34:* 183–210.

Schürmann, F.W. (1995) Common and special features of the nervous system of Onychophora: A comparison with Arthropoda, Annelida, and some other invertebrates. *In* The Nervous System of Invertebrates: An Evolutionary and Comparative Approach (ed. by O. Breidbach, and W. Kutsch), Birkhäuser, Basel, pp. 139–158.

Schwenk, K. (1994/95) A utilitarian approach to evolutionary constraint. Zoology-Anal. Complex Systems, *98:* 251–262.

Selbie, L.A., and S.J. Hill (1998) G protein-coupled-receptor cross-talk: The fine-tuning of multiple receptor-signalling pathways. Trends Pharmacol. Sci., *19:* 87–93.

Sengupta, P., J.H. Chou, and C.I. Bargmann (1996) *odr-10* encodes a seven transmembrane domain olfactory receptor required for responses to the odorant diacetyl. Cell, *84:* 899–909.

Shepherd, G.M. (1974) The Synaptic Organization of the Brain: An Introduction (1st ed.). Oxford University Press, New York.

Shepherd, G.M. (1992) Modules for molecules. Nature, *358:* 457–458.

Singer, A.G. (1991) A chemistry of mammalian pheromones. J. Steroid Biochem. Mol. Biol., *39:* 627–632.

Singer, A.G., and F. Macrides (1993) Composition of an aphrodisiac pheromone. Chem. Senses, *18:* 630.

Singh, R.N., and S.V. Nayak (1985) Fine structure and primary sensory projections of sensilla on the maxillary palp of *Drosophila melanogaster* Meigen (Diptera: Drosophilidae). Int. J. Insect. Morphol., *14:* 291–306.

Sklar, P.B., R.R.H. Anholt, and S.H. Snyder (1986) The odorant-sensitive adenylate cyclase of olfactory receptor cells. J. Biol. Chem., *261:* 15538–15543.

Speca, D.J., D.M. Lin, P.W. Sorensen, E.Y. Isacoff, J. Ngai, and A.H. Dittman (1999) Functional identification of a goldfish odorant receptor. Neuron, *23:* 487–498.

Steinbrecht, R.A. (1996) Are odorant-binding proteins involved in odorant discrimination? Chem. Senses, *21:* 719–727.

Stengl, M. (1994) Inositol-trisphosphate-dependent calcium currents precede cation currents in insect olfactory receptor neurons *in vitro*. J. Comp. Physiol. A, *174:* 187–194.

Steullet, P., O. Dudar, T. Flavus, M. Zhou, and C.D. Derby (2001) Selective ablation of antennular sensilla on the Caribbean spiny lobster *Panulirus argus* suggests that dual antennular chemosensory pathways mediate odorant activation of searching and localization of food. J. Exp. Biol., *204:* 4259–4569.

Stocker, R.F., R.N. Singh, M. Schorderet, and O. Siddiqi (1983) Projection patterns of different types of antennal sensilla in the antennal glomeruli of *Drosophila melanogaster*. Cell Tissue Res., *232:* 237–248.

Störtkuhl, K.F., and R. Kettler (2001) Functional analysis of an olfactory receptor in *Drosophila melanogaster*. Proc. Natl. Acad. Sci. USA, *98:* 9381–9385.

Strausfeld, N.J. (1998) Crustacean-insect relationships: The use of brain characters to derive phylogeny amongst segmented invertebrates. Brain Behav. Evol., *52:* 186–206.

Strausfeld, N.J., and J.G. Hildebrand (1999) Olfactory systems: Common design, uncommon origins? Curr. Opin. Neurobiol., *9:* 634–639.

Strausfeld, N.J., E.K. Buschbeck, and R.S. Gomez (1995) The arthropod mushroom body: Its functional roles, evolutionary enigmas and mistaken identities. *In* The Nervous System of Invertebrates: An Evolutionary and Comparative Approach (ed. by O. Breidbach and W. Kutsch), Birkhäuser, Basel, pp. 349–381.

Strausfeld, N.J., L. Hansen, Y. Li, R.S. Gomez, and K. Ito (1998) Evolution, discovery, and interpretations of arthropod mushroom bodies. Learn. Mem., *5:* 11–37.

Strotmann, J., I. Wanner, T. Helfrich, A. Beck, and H. Breer (1994) Rostro-caudal patterning of receptor-expressing olfactory neurones in the rat nasal cavity. Cell Tissue Res., *278:* 11–20.

Tegoni, M., P. Pelosi, F. Vincent, S. Spinelli, V. Campanacci, S. Grolli, R. Ramoni, and C. Cambillau (2000) Mammalian odorant binding proteins. Biochim. Biophys. Acta, *1482:* 229–240.

Treloar, H., E. Walters, F. Margolis, and B. Key (1996) Olfactory glomeruli are innervated by more than one distinct subset of primary sensory olfactory neurons in mice. J. Comp. Neurol., *367:* 550–562.

Treloar, H.B., A.L. Purcell, and C.A. Greer (1999) Glomerular formation in the developing rat olfactory bulb. J. Comp. Neurol., *413:* 289–304.

Troemel, E.R., J.H. Chou, N.D. Dwyer, H.A. Colbert, and C.I. Bargmann (1995) Divergent seven transmembrane receptors are candidate chemosensory receptors in C. elegans. Cell, *83:* 207–218.

Utsumi, M., K. Ohno, Y. Kawasaki, M. Tamura, T. Kubo, and M. Tohyama (1999) Expression of major urinary protein genes in the nasal glands associated with general olfaction. J. Neurobiol., *39:* 227–236.

Valverde, F., M. Santacana, and M. Heredia (1992) Formation of an olfactory glomerulus: Morphological aspects of development and organization. Neuroscience, *49:* 255–275.

Vassar, R., S.K. Chao, R. Sitcheran, J.M. Nunez, L.B. Vosshall, and R. Axel (1994) Topographic organization of sensory projections to the olfactory bulb. Cell, *79:* 981–991.

Vassar, R., J. Ngai, and R. Axel (1993) Spatial segregation of odorant receptor expression in the mammalian olfactory epithelium. Cell, *74:* 309–318.

Vickers, N.J., T.A. Christensen, and J.G. Hildebrand (1998) Combinatorial odor discrimination in the brain: Attractive and antagonist odor blends are represented in distinct combinations of uniquely identifiable glomeruli. J. Comp. Neurol., *400:* 35–56.

Vogt, R.G. (1987) The molecular basis of pheromone reception: Its influence on behavior. *In* Pheromone Biochemistry (ed. by G.D. Prestwich and G.L. Blomquist), Academic Press, New York, pp. 385–431.

Vogt, R.G., and L.M. Riddiford (1981) Pheromone binding and inactivation by moth antennae. Nature, *293:* 161–163.

Vogt, R.G., F.E. Callahan, M.E. Rogers, and J.C. Dickens (1999) Odorant binding protein diversity and distribution among the insect orders, as indicated by LAP, an OBP-related protein of the true bug *Lygus lineolaris* (Hemiptera, Heteroptera). Chem. Senses, *24:* 481–495.

Vogt, R.G., G.D. Prestwich, and M.R. Lerner (1991) Odorant-binding-protein subfamilies associate with distinct classes of olfactory receptor neurons in insects. J. Neurobiol., *22:* 74–84.

Vogt, R.G., L.M. Riddiford, and G.D. Prestwich (1985) Kinetic properties of a sex pheromone-degrading enzyme: The sensillar esterase of *Antheraea polyphemus*. Proc. Natl. Acad. Sci. USA, *82:* 8827–8831.

Vorontsova, M.N., L.P. Nezlin, and I.A. Meinertzhagen (1997) Nervous system of the larva of the ascidian *Molgula citrina* (Alder and Hancock, 1848). Acta Zool. Stockholm, *78:* 177–185.

Vosshall, L.B. (2001) The molecular logic of olfaction in *Drosophila*. Chem. Senses, *26:* 207–213.

Vosshall, L.B., H. Amrein, P.S. Morozov, A. Rzhetsky, and R. Axel (1999) A spatial map of olfactory receptor expression in the *Drosophila* antenna. Cell, *96:* 725–736.

Vosshall, L.B., A.M. Wong, and R. Axel (2000) An olfactory sensory map in the fly brain. Cell, *102:* 147–159.

Wachowiak, M., and L.B. Cohen (1999) Presynaptic inhibition of primary olfactory afferents mediated by different mechanisms in lobster and turtle. J. Neurosci., *19:* 8808–8817.

Wachowiak, M., and L.B. Cohen (2001) Representation of odorants by receptor neuron input to the mouse olfactory bulb. Neuron, *32:* 723–735.

Wake, D.B. (1991) Homoplasy: The result of natural selection, or evidence of design limitations? Am. Nat., *138:* 543–567.

Weth, F., W. Nadler, and S. Korsching (1996) Nested expression domains for odorant receptors in zebrafish olfactory epithelium. Proc. Natl. Acad. Sci. USA, *93:* 13321–13326.

White, J.G., E. Southgate, J.N. Thomson, and S. Brenner (1986) The structure of the nervous system of the nematode *Caenorhabditis elegans*. Phil. Trans. Royal Soc. B, *314:* 1–340.

Wicht, H., and R.G. Northcutt (1992) The forebrain of the Pacific hagfish: A cladistic reconstruction of the ancestral craniate forebrain. Brain Behav. Evol., *40:* 25–64.

Wong, S.T., K. Trinh, B. Hacker, G.C. Chan, G. Lowe, A. Gaggar, Z. Xia, G.H. Gold, and D.R. Storm (2000) Disruption of the type III adenylyl cyclase gene leads to peripheral and behavioral anosmia in transgenic mice. Neuron, *27:* 487–497.

Wray, G.A. (2002) Do convergent developmental mechanisms underlie convergent phenotypes? Brain Behav. Evol., *59:* 327–336.

Young, J.Z. (1965) The central nervous system of *Nautilus.* Phil. Trans. R. Soc. Lond. B, *249:* 1–25.

Young, J.Z. (1971) The Anatomy of the Nervous System of *Octopus vulgaris.* Clarendon, Oxford.

Zakon, H.H. (2002) Convergent evolution on the molecular level. Brain Behav. Evol., *59:* 250–261.

Zhainazarov, A.B., and B.W. Ache (1995a) Odor-induced currents in *Xenopus* olfactory receptor cells measured with perforated-patch recording. J. Neurophysiol., *74:* 479–483.

Zhainazarov, A.B., and B.W. Ache (1995b) Na^+-activated nonselective cation channels in primary olfactory neurons. J. Neurophysiol., *73:* 1774–1781.

Zhainazarov, A.B., and B.W. Ache (1997) Gating and conduction properties of a sodium-activated cation channel from lobster olfactory receptor neurons. J. Membr. Biol., *156:* 173–190.

Zhainazarov, A.B., R. Doolin, J.D. Herlihy, and B.W. Ache (2001) Odor-stimulated phosphatidylinositol 3-kinase in lobster olfactory receptor cells. J. Neurophysiol., *85:* 2537–2544.

Zhainazarov, A.B., R.E. Doolin, and B.W. Ache (1998) Sodium-gated cation channel implicated in the activation of lobster olfactory receptor neurons. J. Neurophysiol., *79:* 1349–1359.

Zhang, Y., J.H. Chou, J. Bradley, C.I. Bargmann, and K. Zinn (1997) The *Caenorhabditis elegans* seven-transmembrane protein ODR-10 functions as an odorant receptor in mammalian cells. Proc. Natl. Acad. Sci. USA, *94:* 12162–12167.

Zhao, H., L. Ivic, J.M. Otaki, M. Hashimoto, K. Mikoshiba, and S. Firestein (1998) Functional expression of a mammalian odorant receptor. Science, *279:* 237–242.

Brain, Behavior and Evolution

Brain Behav Evol 2002;59:294–311

Evolutionary Convergence and Shared Computational Principles in the Auditory System

C.E. Carr D. Soares

Department of Biology, University of Maryland, College Park, Md., USA

Key Words

Birds · Mammals · Ontogeny · Time coding · Evolution

Abstract

Precise temporal coding is a hallmark of the auditory system. Selective pressures to improve accuracy or encode more rapid changes have produced a suite of convergent physiological and morphological features that contribute to temporal coding. Comparative studies of temporal coding also point to shared computational strategies, and suggest how selection acts to improve coding. Both the avian cochlear nucleus angularis and the mammalian cochlear nuclei have heterogeneous cell populations, and similar responses to sound. These shared characteristics may represent convergent responses to similar selective pressures to encode features of airborne sound.

Introduction

In the auditory system, precise encoding of temporal information has direct behavioral relevance for sound localization and communication [Heffner and Heffner, 1992; Hafter and Trahiotis, 1997]. One might therefore expect selection to improve accuracy or encode the more rapid changes associated with sensitivity to higher frequency sounds. In fact, similar physiological and morphological features that might improve temporal coding characterize the auditory brainstem of both birds and mammals. We will make two points about the evolution of the amniote auditory brainstem in this review. First, the similarities among brainstem circuits that encode sound in birds and mammals may be the result of convergence. Second, the existence of convergent circuits allows us to identify algorithms shared by the auditory system of birds and mammals, and to argue that these are suited to extracting the stimulus variables relevant for auditory coding; thus, studies of evolution can inform computational neurobiology.

The fossil record provides evidence that some of the similarities in auditory structures in birds and mammals may be the result of convergence. Tetrapod tympanic ears are not homologs, and recent work has shown that tympanic ears may have evolved independently at least five times, in synapsids, lepidosauromorph diapsids, archosauromorph diapsids, probably turtles, and amphibians [Clack, 1997]. Wilczynski [1984] has argued that these peripheral changes would have different reorganizing effects upon the ancestral population of brainstem auditory

KARGER

Fax + 41 61 306 12 34
E-Mail karger@karger.ch
www.karger.com

Accessible online at:
www.karger.com/journals/bbe

C.E. Carr
Department of Biology
University of Maryland
College Park, MD 20742 (USA)
Tel. +1 301 405 6915, Fax +1 301 314 9358, E-Mail cc117@umail.umd.edu

neurons, leading to the parallel evolution of the central targets of the auditory nerve. Further developments in ancestral mammals, such as moveable ears and multiple ossicles, might have had additional reorganizing effects. In this paper, we will argue that similar features in birds and mammals represent convergence.

Convergence is a plausible outcome of auditory system evolution, because animals with tympanic ears should experience similar constraints in detecting sounds. Indeed, the essential features of auditory coding, and temporal coding in particular, are very similar in birds and mammals. Understanding the evolutionary and developmental events behind the similar form and function of temporal coding cells in birds and mammals will require a detailed knowledge of multiple species under study, and of their phylogenetic relationships. This requires deliberate concentration on the differences among animals. When we know more about the neural circuits in different groups, we can construct scenarios about how they evolved. In this review we compare similar coding strategies in the time coding neurons of birds and mammals, and use these comparative studies to argue that natural selection has produced suitable, convergent solutions to the problems of temporal coding (sections 'Encoding Temporal Information' and 'Coincidence Detection and ITD Coding', below). In the section 'Encoding Sound: ...', we will compare avian and mammalian auditory brainstem processing of other aspects of the auditory stimulus in order to identify shared features.

Encoding Temporal Information

Barn owls are able to catch mice on the basis of auditory cues alone [Konishi, 1973]. Accurate and precise processing of the auditory stimulus is required for this behavior. Auditory nerve fibers encode temporal information by phase-locking to the waveform of the acoustic stimulus, and this temporal information is preserved in projections to the primary auditory nuclei, nucleus magnocellularis (NM), and nucleus angularis (NA). Three lines of evidence show that accurate temporal coding is important. First, measurements of the vector strength of the auditory nerve signal (calculated from the variability in the timing of action potentials with respect to the phase of the acoustic stimulus) show an improvement in high frequency phase-locking in the owl as compared to other animals by an octave or more [Köppl, 1997]. Second, models of coincidence detection perform better when the vector strength of the inputs improves [Colburn et al., 1990; Simon et al., 1999]. Third, inactivation of NM neurons with lidocaine removes sensitivity to interaural time differences (ITDs) from the responses of midbrain space-mapped neurons [Takahashi et al., 1984]. We will review the features associated with preserving temporal cues up to the point where ITDs are detected in the nucleus laminaris (NL).

There are several shared features of temporal coding circuits in birds and mammals. These include high quality inputs, presynaptic specializations to make neurotransmitter release both precise and modifiable, and postsynaptic specializations, including specific glutamate receptors, potassium conductances and characteristic neuronal morphology.

Precise Synaptic Transmission

The task of accurately representing the stimulus phase becomes more difficult with increasing stimulus frequency because temporal dispersion decreases [Hill et al., 1989]. This is because the absolute temporal precision required for phase-locking to high frequencies is greater than that needed for low frequencies, i.e., the same variation in temporal jitter of spikes translates to greater variation in terms of degrees of phase for high frequencies. Hill et al. [1989] estimated phase-locking in the auditory fibers of the pigeon in terms of the commonly used synchronicity index (vector strength) as well as by measuring temporal dispersion. Vector strength of phase-locking decreased for frequencies above 1 kHz. Temporal dispersion, however, also decreased with frequency, indicating enhanced temporal synchrony as frequency increased [Köppl, 1997]. The upper frequency limit of phase-locking therefore appears to depend primarily upon the ability of auditory hair cells to encode stimulus phase, and on irreducible jitter in the timing of spikes [Carr and Amagai, 1996]. It is about 8 kHz for barn owls, and between 4–6 kHz for most other birds and mammals studied [Köppl, 1997].

Endbulb terminals may have emerged as an adaptation for accurate transmission of phase information for frequencies above 500 Hz, perhaps associated with the development of hearing in land vertebrates (fig. 1). Large somatic terminals have been found in all amniote groups examined. There are no data on turtles, but large somatic terminals and smaller boutons are found in all NM of the alligator lizard [Szpir et al., 1990]. In crocodilian NM, the rostral high best frequency NM neurons receive endbulb-like projections, whereas lower best frequency NM neurons receive bouton terminals [Soares, unpublished]. Elaborate endbulb terminals are found in both birds and mammals [Ryugo, 1991]. Thus it is possible that a large somatic terminal may have been present in the amniote common

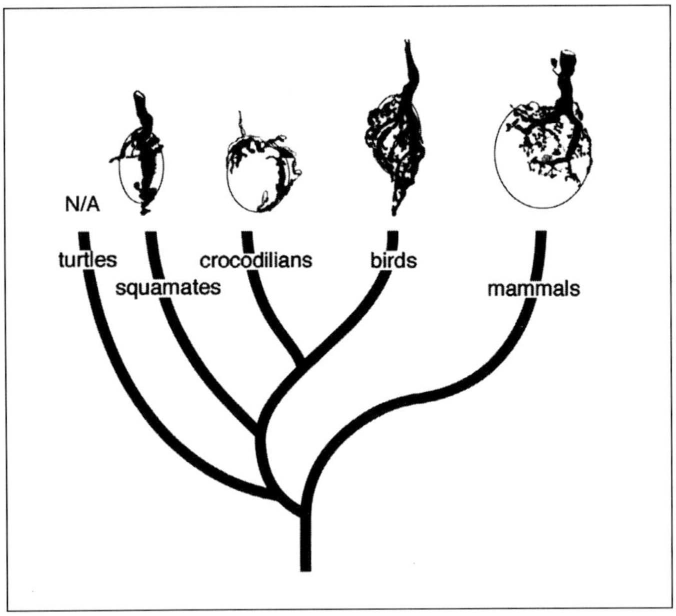

Fig. 1. Large somatic terminals appear in NM of birds, crocodilians and in the alligator lizard, as well as the mammalian ventral cochlear nucleus. These terminals are similarly elaborate in birds and mammals, and fractal analysis reveals similar morphological complexity [Ryugo et al., 1996; Carr et al., 1997]. This complexity appears to have developed in parallel in both birds and mammals. For example, although the terminals are similar, the mechanisms by which transmitter release is modulated are not the same. Activation of presynaptic glycine receptors enhances transmitter release in mammalian MNTB endbulbs by triggering a weakly depolarizing Cl^- current in the terminal. This depolarization enhances release by activating Ca^{2+} channels and increasing resting intraterminal Ca^{2+} concentrations [Brenowitz and Trussell, 2001]. In birds, activation of presynaptic $GABA_B$ receptors affects the endbulb to NM synapse by minimizing AMPA receptor desensitization and therefore enhancing synaptic strength during high-frequency activity [Turecek and Trussell, 2000].

ancestor, and have developed in parallel in lizards, archosaurs, and mammals to mediate accurate transmission of temporal information at higher sound frequencies. Support for the hypothesis that endbulbs developed to facilitate transmission of high best frequency phase-locking comes from comparisons of low and high best frequency regions of NM. Endbulb terminals are not essential for transmission of phase-locked spikes at low frequencies. The very low best frequency cells of the NM receive large bouton terminals from the auditory nerve and phase-lock to frequencies below ~1 kHz [Köppl, 1997], whereas in crocodilian NM, only the rostral high best frequency NM neurons receive endbulb-like projections [Soares, unpublished].

The advantages of the endbulb are that synaptic currents are injected into the cell body, not the dendrites. Furthermore, the invasion of the presynaptic action potential into the calyx leads to the synchronous release of quanta at many endbulb release sites, giving this synapse a high safety factor of transmission [Isaacson and Walmsley, 1996]. The invading presynaptic action potential is extremely narrow (<200 μs at room temperature in 14-day-old rats and ~250 μs at 35 °C in postnatal day 8–10 animals) [Borst et al., 1995; Taschenberger and von Gersdorff, 2000] probably due to rapid repolarization mediated by specific potassium conductances. Calcium influx into the presynaptic terminal is also brief and occurs only during the falling phase of the presynaptic action potential [Borst and Sakmann, 1996]. Because the action potential is narrow, its down stroke occurs quickly, as does calcium influx, reducing the synaptic delay. In addition, the brief period of calcium influx produces a confined and phasic period of neurotransmitter release which also increases the temporal precision of transmission across the synapse [Sabatini and Regehr, 1999].

Large Neurons

The requirement of temporal accuracy at the synapse, particularly for high frequency inputs, may have driven the evolution of pathways that process precise temporal information. Improvement in intrinsic accuracy of neurons can be achieved through anatomical and physiological specializations of both presynaptic and postsynaptic structures that maximize the signal while minimizing the noise. One general strategy is to make everything large. Larger somata and axons are less vulnerable to noise caused by stray currents, as their low input resistance and large current generating ability would keep the influence of voltage fluctuations to a minimum. Many of the known time-coding pathways include large cells [Carr and Amagai, 1996].

Enlarged size must be accompanied by an increase in synaptic current, as observed (see below). Further, the currents also should have a fast rise time to minimize the influence of ambient voltage fluctuations on the timing of spikes. One solution is to have large terminals that partially engulf the postsynaptic cell, presumably translating into massive release of neurotransmitter [Zhang and Trussell, 1994; Trussell, 1997]. Fast rise times can be enhanced by reducing the electrotonic distance between the synapse and the site of integration, minimizing the attenuation of synaptic current. This occurs in time-coding electric fish neurons and in the cells of the nucleus magnocellularis and the nucleus laminaris in birds [Smith and Rubel, 1979; Jhaveri and Morest, 1982; Bell and Sza-

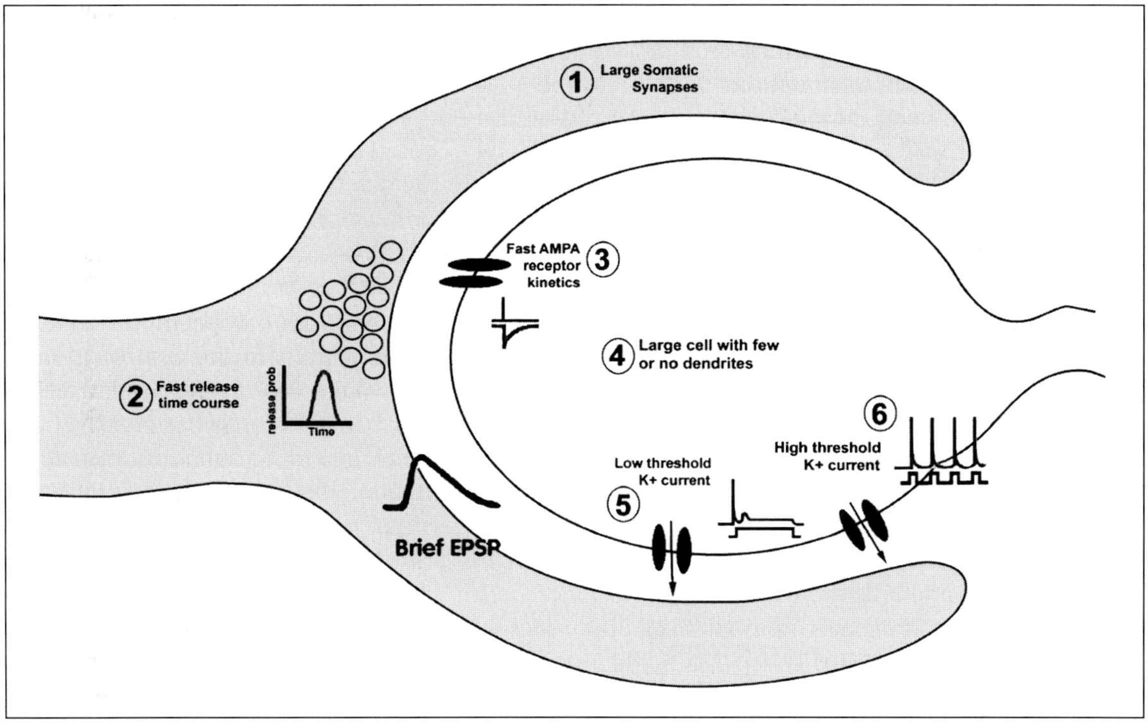

Fig. 2. Summary of physiological and morphological features that characterize time coding neurons. Diagram of endbulb terminal on NM neuron that identifies pre- and postsynaptic specializations including **1** large somatic synapse, **2** fast time course of release, **3** AMPA receptors with rapid kinetics, **4** large postsynaptic cell with few or no dendrites, **5** low threshold K^+ conductances and **6** high threshold K^+ conductances [redrawn from Trussell, 1999].

bo, 1986; Carr and Boudreau, 1993; Kawasaki and Guo, 1996; Amagai et al., 1998].

Behavioral evidence makes it clear that phase-coding systems can extract precise temporal information despite variability (jitter) in the phase-locking of neural spikes to the electrical or auditory stimulus. Large numbers of inputs are often invoked as a mechanism for reducing temporal jitter [Carr, 1986; Kawasaki et al., 1988; Heiligenberg, 1989]. In the simplest model, the averaging of n presynaptic units should reduce the jitter in the postsynaptic neuron by $1/\sqrt{n}$. An important requirement for this convergence to be effective is that the synapse and the spike-generating mechanism in the postsynaptic neuron not themselves contribute significant additional jitter [Carr and Amagai, 1996].

'Auditory' AMPA Receptors

Activation of AMPA type glutamate receptors at endbulb synapses generates brief, large synaptic currents that are suited to the transfer of temporally precise information from pre- to postsynaptic cell [fig. 2; Raman and Trussell, 1992; Zhang and Trussell, 1994]. The brevity of EPSCs in these neurons depends not only on the time course of release but also on the specific properties of the postsynaptic receptors. AMPA receptors are made up of GluR splice variants, and the GluR3 and 4_{flop} isoforms found in time coding auditory neurons have fast kinetics and very rapid desensitization rates, such that the duration of miniature EPSCs in auditory neurons are among the shortest recorded for any neuron [Gardner et al., 1999; Trussell, 1999]. These fortunate kinetics result from a characteristic 'auditory' pattern of expression [Parks, 2000]. In the chicken NM, where a homogeneous population of neurons makes this possible, mRNA analysis has revealed the relative abundance of the four AMPA receptor subunits [Ravindranathan et al., 1996]. These receptors have very low levels of GluR2, and higher levels of the 'fast' flop splice variants of GluR 3 and 4 [Ravindranathan et al., 1996; Parks, 2000]. Similar splice variants characterize mammalian bushy cells [Gardner et al., 2001].

Low and High Threshold Potassium Conductances

Although brief EPSCs underlie the rapid synaptic potential changes seen in time coding neurons, the intrinsic electrical properties of these neurons also shape the synaptic response as well as the temporal firing pattern. Of particular interest are the voltage sensitive K^+ conductances. The importance of these conductances in sculpting the response properties of auditory neurons was first demonstrated by Manis and Marx [1991] who showed that differences in the electrical responses of bushy cells and stellate cells in the mammalian cochlear nucleus can be attributed to a distinct complement of outward K^+ currents in each cell type. At least two K^+ conductances underlie phase-locked responses in auditory neurons: a low threshold conductance (LTC) and a high threshold conductance (HTC) [fig. 2; Manis and Marx, 1991; Reyes et al., 1994; Brew and Forsythe, 1995; Rathouz and Trussell, 1998; Wang et al., 1998].

The LTC activates at potentials near rest and is largely responsible for the outward rectification and non-linear current voltage relationship around the resting potential seen in a number of auditory neurons [for review, see Oertel, 1999, and fig. 5]. Activation of the LTC leads to a short active time constant so that the effects of excitation are brief and do not summate in time [Wu and Oertel, 1984]. Only large EPSPs reaching threshold before significant activation of the LTC would produce spikes with short latencies, whereas small EPSPs that depolarize the membrane more slowly would allow time for LTC activation to shunt the synaptic current and prevent action potential generation and thus long latency action potentials. Blocking the LTC elicits multiple spiking in response to depolarizing current injection [Manis and Marx, 1991; Rathouz and Trussell, 1998] or synaptic activation [Brew and Forsythe, 1995]. It is believed that K^+ channels underlying the LTC are composed of Kv1.1 and Kv1.2 subunits. Both subunits are expressed in auditory neurons, although the subcellular distribution is unknown [Grigg et al., 2000].

The HTC is characterized by an activation threshold around –20 mV and fast kinetics [Brew and Forsythe, 1995; Rathouz and Trussell, 1998; Wang et al., 1998]. These features of the HTC result in fast spike repolarization and a large but brief after-hyperpolarization without influencing input resistance, threshold, or action potential rise time. Thus, the HTC can keep action potentials brief without affecting action potential generation. In addition, the HTC minimizes Na^+ channel inactivation, allowing cells to reach firing threshold sooner and thereby facilitating high frequency firing. Relatively specific pharmacological blockade of the HTC broadens action potentials and reduces the fast after-hyperpolarization [Brew and Forsythe, 1995]. Furthermore, blockade of the HTC diminishes the ability of the medial nucleus of the trapezoid body (MNTB) neurons to follow high frequency stimuli in the range of 300–400 Hz, but has little effect on responses to low frequency (<200 Hz) stimulation [Wang et al., 1998].

Currents produced by Kv3 channels share many characteristics of the HTC, including a positive activation range, rapid deactivation kinetics, and pharmacological sensitivity, and most likely underlie the HTC. Neurons that fire fast, including many auditory neurons, express high levels of Kv3 mRNA and protein [Perney and Kaczmarek, 1997; Li et al., 2001; Parameshwaran et al., 2001]. Interestingly, in several auditory nuclei including avian NM and nucleus laminaris (NL) [Parameshwaran et al., 2001], and rat MNTB [Li et al., 2001], Kv3.1 protein expression varies along the tonotopic map such that mid to high best frequency neurons are most strongly immunopositive, whereas neurons with very low best frequencies are only weakly immunopositive. A high to low frequency gradient of Kv3.3 expression has also been observed in electrosensory lateral line lobes of a weakly electric fish [Rashid et al., 2001]. These results suggest that the electrical properties of higher order auditory neurons may vary with frequency tuning. Because no differences in either spontaneous or driven rates have been observed across the tonotopic axis, however, Kv3 channels may be functioning as more than just a facilitator of high frequency firing, and may also enhance the temporal precision of spike discharges.

Distribution of Kv3.1 protein in auditory neurons is largely somatic and/or axonal, consistent with its role in spike repolarization [Perney and Kaczmarek, 1997; Li et al., 2001; Parameshwaran et al., 2001]. EM studies have shown that Kv3.1 is present in the membranes of endbulb terminals onto MNTB neurons, suggesting that Kv3.1 channels may be at least partially responsible for the extremely brief action potential seen at this terminal. Kv3.1 protein is also present in the NM axons innervating NL in the owl, but not in the chicken [Parameshwaran et al., 2001]. The increased levels of HTC associated with Kv3.1 expression in owl NM axons would reduce the width of the action potential invading the NM terminals and thus the amount of neurotransmitter released. Modeling of coincidence detector neurons suggests that an increase in the width of the input EPSC could impair ITD coding [Simon et al., 1999; see below]. Thus, the selective increase of Kv3.1-like currents in the NM delay line axons in owl might contribute to the temporal synchrony necessary for accurate phase-locking.

There are differences among various temporal coding circuits. Examples from potassium channel expression include the coincidence detector neurons in birds and mammals. Neurons of the medial superior olive (MSO) do not express either Kv3.1 mRNA or protein, unlike NL neurons. They do, however, express high levels of Kv3.3 message [Perney and Kaczmarek, 1997; Grigg et al., 2000]. Thus, differences in Kv3 expression between NL and MSO structures may reflect species differences in the expression of Kv3 subfamily members, or this variation might represent a significant physiological difference. Other examples are the differences in the subcellular localization of Kv3.1 protein in chick and owl NM axons described above. In both the owl and the chicken, Kv3.1 is targeted postsynaptically in NM and NL. The major difference in localization of Kv3.1 protein between the two birds is that Kv3.1 protein is expressed in the NM axons and terminals that act as delay lines in the barn owl NL. The presynaptic localization of Kv3.1 in owl might be a specialization enabling neurons in owl NM to transmit high frequency temporal information with little jitter. Nishikawa [1997] has proposed that these types of interspecific differences could occur as small changes superimposed upon a basic developmental plan.

Similar potassium conductances characterize other time coding cells. There are numerous examples, many discussed in Oertel's 1999 review. In addition to the NM and mammalian cochlear nucleus bushy cells discussed above, the coincidence detectors in the avian NL and mammalian medial superior olive also express similar conductances and respond with temporal precision to the auditory stimulus (see below). The reasons for temporal precision are clear for the circuit that detects interaural time differences (ITDs). There are also other aspects of the auditory stimulus that require temporal precision. In particular, the mammalian cochlear nucleus octopus cells form the origin of a circuit that encodes timing of events, especially broadband transients. Octopus cells produce the briefest, most sharply timed synaptic responses in the mouse cochlear nucleus [Golding et al., 1995]. Octopus cells are characterized by both a large low threshold conductance and a high threshold conductance [Bal and Oertel, 2000]. Type II cells in the ventral nucleus of the lateral lemniscus produce sharply timed responses and receive endbulb input from octopus cells [Wu, 1999]. Thus selection for temporal accuracy might in each case drive the expression of conductances that improve neuronal performance and behavioral accuracy.

Coincidence Detection and ITD Coding

In birds and mammals, precisely timed spikes encode the timing of acoustic stimuli, and interaural acoustic disparities propagate to binaural processing centers such as the avian NL and the mammalian medial superior olive (MSO) [Young and Rubel, 1983; Carr and Konishi, 1990; Joris et al., 1998; fig. 3]. The projections from the NM to NL and from mammalian spherical bushy cells to MSO resemble the Jeffress model for encoding ITD [Jeffress, 1948]. The Jeffress model describes a circuit composed of two elements: delay lines and coincidence detectors. The coincidence detectors are arranged in an array, every element of which has a different relative delay between its ipsilateral and contralateral excitatory inputs. Thus, ITD is encoded into the position (a place code) of the coincidence detector whose delay lines best cancel out the acoustic ITD [for reviews, see Konishi, 1991; Joris et al., 1998]. Neurons of NL and MSO phase-lock to both monaural and binaural stimuli but respond maximally when phase-locked spikes from each side arrive simultaneously, i.e., when the difference in the conduction delays compensates for the ITD [Goldberg and Brown, 1969; Carr and Konishi, 1990; Yin and Chan, 1990; Overholt et al., 1992; Peña et al., 1996].

Models of Coincidence Detection Relate Dendritic Structure to Detection of Interaural Time Differences

A singular feature of the coincidence detectors in mammals, and of low best frequency NL cells in birds, is their common morphological organization. Both are bitufted neurons with inputs from each ear segregated on the dendrites (fig. 4). Modeling studies suggest this dendritic organization improves coincidence detection [Agmon-Snir et al., 1998]. Thus the cell morphology and the spatial distribution of the inputs enrich the computational power of these neurons beyond that expected from 'point neurons'. How does the dendritic structure of the coincidence detectors enhance their computational ability? An ITD discriminator neuron should fire when inputs from two independent neural sources coincide (or almost coincide), but not when two inputs from the same neural source (almost) coincide. A neuron that sums its inputs linearly would not be able to distinguish between these two scenarios. To understand this mechanism, we constructed a biophysically detailed model of coincidence detector neurons using NEURON [Simon et al., 1999].

Two dendritic non-linearities aid coincidence detection. First, synaptic inputs arriving at the same dendritic

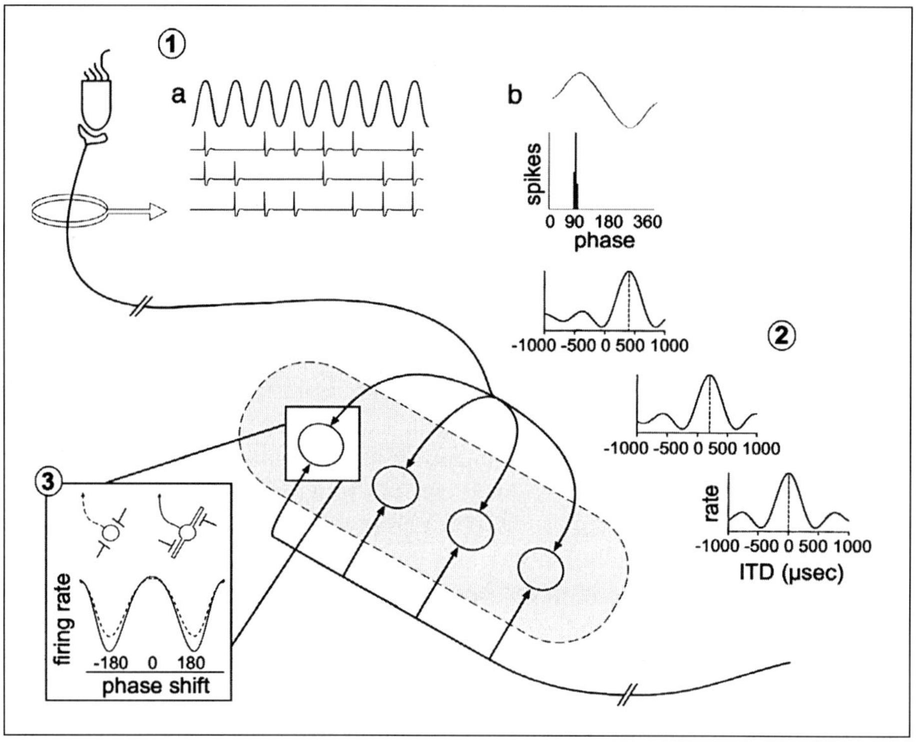

Fig. 3. Summary of ITD detection circuits. **1** Temporal code is maintained from hair cell to cochlear nucleus to site of ITD detection. **a** Phase-locking, with multiple input channels all encoding the phase of the auditory signal. **b** Period histogram whereby the timing of the spikes in **a** are plotted with respect to stimulus phase. **2** Place code for ITD achieved by Jeffress-like circuit. Monaural channels act as delay lines and project to an array of coincidence detectors that each tap the signal at a different ITD. The cells for which the internal (axonal) delay is equal but opposite to the acoustic ITD are maximally active [from Joris et al., 1998]. The delay lines create a map of ITD whereby the temporal code is transformed into a place code. The best evidence for this is in the cat [for review, see Joris et al., 1998]. **3** Coincidence detectors respond maximally to simultaneous inputs from each ear. Their firing rate drops to a minimum when these inputs are 180° out of phase. Models of coincidence detectors suggest that segregation of inputs from each ear onto bitufted dendrites improves coincidence detection by reducing the possibility of a response to a non-synchronous input [Agmon-Snir et al., 1998].

compartment sum non-linearly because the driving force decreases with depolarization [Agmon-Snir et al., 1998]. Hence, the net synaptic current from several inputs arriving simultaneously at nearby sites on the same dendrite is smaller than the net current generated if these inputs are distributed on different dendrites. As a result, the conductance threshold, or minimum synaptic conductance needed to trigger a somatic action potential, is higher when the synaptic events are on the same dendrite, compared to when they are split between the bipolar dendrites. Second, each dendrite acts as a current sink for inputs on the other dendrite, consequently increasing the voltage change needed to trigger a spike at the soma when inputs arrive only on one side. This effect is boosted by the presence of a low threshold K^+ conductance similar to that found in NM and bushy neurons, so that out of phase inputs are subtractively inhibited [Simon et al., 2000]. With only monaural input, the LTC in the opposite dendrite is somewhat activated, producing a mild current sink. However, when there are recent EPSPs in the opposite dendrite due to out-of-phase inputs, the LTC is strongly activated and acts as a large current sink suppressing spike initiation. Thus, the model predicts the experimental finding [Goldberg and Brown, 1969; Carr and Konishi, 1990; Yin and Chan, 1990] that the monaural firing rate, although lower than the binaural in-phase rate, is higher than the binaural out-of phase rate.

One dendritic effect diminishes with increasing stimulus frequency. When typical chick-like parameters are used, sub-linear summation in the dendrites only improves coincidence detection below 2 kHz, after which discrimination between in-phase and out-of-phase inputs is poor [Agmon-Snir et al., 1998]. This is consistent with observations from rabbit MSO neurons, where ITD sensitivity has only been observed for sounds at or below 2 kHz [Batra et al., 1997]. The second dendritic non-linearity, subtractive inhibition of out-of-phase inputs, improves coincidence detection at all frequencies [Simon et al., 2000] and might therefore be most significant in avian coincidence detectors between 2–8 kHz. It is also clear that the quality of phase-locked inputs has some bearing on coincidence detection: typical chick-like parameters but with barn owl-like phase-locking allow ITD discrimination up to 4–6 kHz [Simon et al., 2000]. The benefits conveyed by the neuronal structure of the coincidence detectors allow us to argue that selective forces have directed the convergent evolution of coincidence detectors in the bird NL and mammalian MSO.

Convergent Aspects of ITD Coding in Amniote Auditory Systems

ITD detection is a common feature of avian and mammalian auditory systems, although the resolving ability of the discrimination task varies among different species [Heffner, 1997]. The ability of the auditory system to use ITD cues to localize sounds requires accurate encoding of temporal information. An important feature of both avian and mammalian coincidence detectors is that they share physiological features with time coding neurons in the cochlear nuclei. Coincidence detectors exhibit specific K^+ conductances that lead to a single or few well-timed spikes in response to a depolarizing stimulus in vitro. The

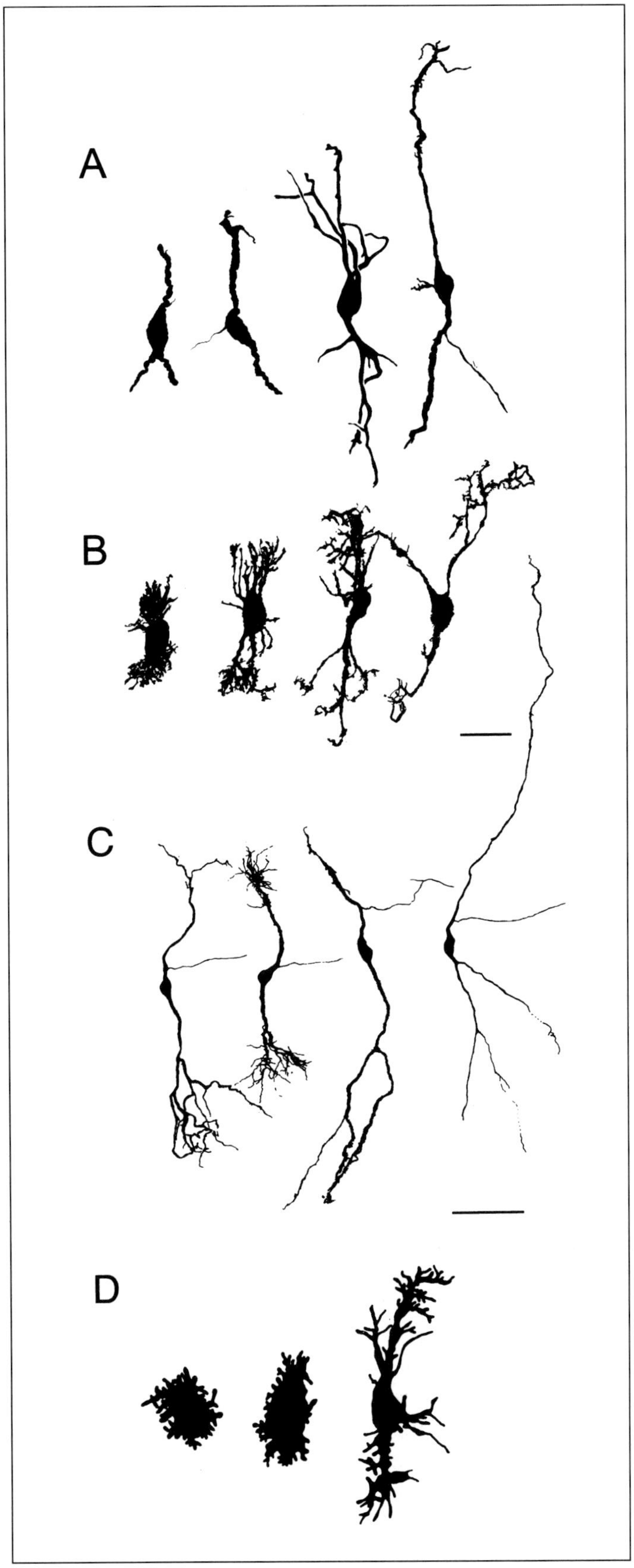

Fig. 4. Coincidence detectors share bitufted morphology. **A** Alligator NL neurons labeled with Golgi technique, from presumed high to low best frequency regions of NL [left to right; Soares, unpublished]. **B** Chicken NL neurons labeled with Golgi technique, from high to low best frequency regions of NL [Jhaveri and Morest, 1982]. Bar for **A** and **B** = 40 μm, for **D** = 20 μm. **C** Guinea pig MSO neurons [Smith, 1995]. Bar = 100 μm. **D** Barn owl NL labeled with Golgi technique [Carr et al., 1997]. Dendritic length increases from left to right except in the principal cells of the medial superior olive (MSO) from the guinea pig, where a frequency gradient is not apparent [adapted from Smith, 1995]. The bipolar architecture and the segregation of the inputs arriving from both ears is common to both mammalian and avian coincidence detectors. In the barn owl, coincidence detectors have lost this bipolar organization (except in low best frequency regions), and their short dendrites radiate around the cell body.

LTC should decrease the effective membrane time constant. Modeling studies suggest that these fast conductances are instrumental in keeping the firing rate near zero when the inputs are completely out of phase while allowing non-zero firing rate when the inputs are monaural [Agmon-Snir et al., 1998; Simon et al., 1999].

Both avian and mammalian coincidence detectors share physiological features with NM neurons and mammalian bushy cells. Coincidence detectors exhibit specific K^+ conductances that lead to a single or few well-timed spikes in response to a depolarizing stimulus in vitro [Smith, 1995; Reyes et al., 1996]. The low threshold K^+ channels should decrease the effective membrane time constant; that is, the average membrane time constant for a cell receiving and processing in vivo rates of EPSPs should be shorter than the passive membrane time constant [Mainen and Sejnowski, 1995, 1996; Gerstner et al., 1996]. These fast conductances may be critical to coincidence detection.

Although coincidence detector neurons in birds and mammals display similar conductances and bipolar morphologies (fig. 4) they are not identical. At the cellular level, mammalian MSO neurons do not express Kv3.1 [Grigg et al., 2000; Li et al., 2001; see section on 'Potassium Conductances', above]. They do express high levels of Kv3.3 message [Grigg et al., 2000; Li et al., 2001]. We do not know whether this variation in expression represents a significant physiological difference. With regard to morphology, avian, crocodilian, and mammalian coincidence detectors all have a bitufted dendritic organization, but mammals do not show an appreciable change in dendritic length with best frequency [Cant, 1992; Smith, 1995] (fig. 4). We do not know whether the biophysical or synaptic properties change with best frequency because there are no data comparing the responses of MSO neurons with their best frequency.

At the circuit level, a second substantial difference between bird and mammal coincidence detectors is in inhibitory inputs. In mammals the MSO receives well-timed inhibitory input from the medial and lateral nucleus of the trapezoid body [Cant and Hyson, 1992; Kuwabara and Zook, 1992; Grothe and Sanes, 1994; Kapfer et al., 2002]. In the gerbil, these inhibitory inputs enhance coincidence detection by shifting the peak of the ITD function so as to place the slope within the biological range [Brand et al., 2002]. In birds, inhibitory inputs in NL are more diffuse, and appear to decrease excitability through a gain control mechanism [Funabiki et al., 1998; Yang et al., 1999; Monsivais et al., 2000; Peña et al., 1996].

Because the bitufted structure can improve ITD detection, it is parsimonious to suppose that ITD detection in birds and mammals evolved independently, and that their similarity is due to convergence. Evolutionary and developmental studies will be needed to determine whether or not the coincidence detector neurons in the NL and MSO shared a common bitufted ancestor, or whether this morphology arose independently in the two groups.

Encoding Sound: Similar Strategies in Birds and Mammals

Both birds and mammals encode phase and detect ITD in a similar way. Do they also use similar strategies for encoding the rest of the auditory scene? When comparing the avian NA with the mammalian cochlear nuclei, we have found that both structures have heterogeneous cell populations, and similar responses to sound [Köppl et al., 2001; Soares and Carr, 2001; Soares et al., 2002]. We do not know whether these shared characteristics evolved from heterogeneous ancestral populations, or if they have evolved in parallel as a response to similar selective pressures to encode airborne sound. There are two possible reasons why similarities might be due to convergence. (1) True tympanic ears arose independently in birds and mammals [Clack, 1997; see Introduction]. These peripheral changes would have had different reorganizing effects upon the ancestral population of brainstem auditory neurons. (2) The cell types of the avian and mammalian cochlear nuclei are similar but not identical. We describe these similarities in this section. A satisfactory study will, however, require detailed analyses of the morphology and physiology of cell types in the cochlear nucleus in all amniote groups, including turtles, basal lizards, and crocodilians. We also note that the divisions of the cochlear column in birds and mammals do not appear to be the same. There are two avian cochlear nuclei, with NM containing essentially a single cell type, where as NA has heterogeneous cell types distributed throughout the nucleus. The mammalian cochlear nuclei contain groupings of cells and three adjacent divisions termed dorsal, anteroventral, and posteroventral. We and others have attempted to determine the relationship between NA and the mammalian cochlear nuclei. At present, no scheme can satisfactorily account for the observed similarities and differences.

Nucleus angularis and the Mammalian Dorsal Cochlear Nucleus

There are no true morphological correspondences between NA and the DCN, despite superficial similarity [Boord, 1969; Sachs and Sinnot, 1978; Sachs et al., 1980]. Historically, there were two main reasons for thinking NA and DCN were homologous: (1) the existence of type IV (complex non-monotonic) physiological responses in both NA and DCN [Sachs and Sinnott, 1978; Sachs et al., 1980; for review of DCN see Young et al., 1988]. (2) Boord [1969] proposed that medial NA and NM corresponded to the mammalian ventral cochlear nucleus and that lateral NA was comparable to the DCN. Boord's supposition was supported by the increased cell density and small cell size of lateral NA and by the presence of the lateral afferent tract. These anatomical features created a superficial resemblance to the fusiform and molecular layers of the DCN. We now know that NA is a unitary structure, with the same cell types distributed along the tonotopic axis [Soares and Carr, 2001]. Parsimony would therefore suggest that the type IV responses observed by Sachs and Sinnot [1978] might have emerged in parallel with similar responses in mammalian DCN. The DCN appears to be a unique feature of the mammalian auditory system, associated possibly with the development of motile external ears. Furthermore, unlike the case with the NA, the DCN shares many common features with the cerebellum, including unique cell types and cortical circuitry [Mugnaini et al., 1987, 1997; Wright and Ryugo, 1996]. It can be inferred therefore that at least some parts of DCN have a distinct embryological origin, different than the other cochlear nuclei.

Similar Organization of Multipolar Cell Types with Respect to the Tonotopic Axis in Birds and Mammals

Doucet and Ryugo [1997] described ventral cochlear nucleus neurons that project to the dorsal cochlear nucleus of the rat. They divided these multipolar neurons into two distinct groups: planar and radiate. Planar neurons have dendrites oriented within the isofrequency plane, and Doucet and Ryugo [1997] suggest that these neurons might respond best to a narrow range of frequencies. Radiate multipolar neurons, on the other hand, have long dendrites oriented perpendicular to the isofrequency contours, and might be sensitive to a broader range of frequencies. A third group of multipolar cells is present on the margins of the ventral cochlear nucleus.

Although the radiate neurons in the bird NA do not span as large a portion of the tonotopic axis as the radiate neurons described by Doucet and Ryugo [1997] the basic pattern of planar and radiate cells observed in the rat applies to NA. NA cells can be divided into classes on the basis of their dendritic organization with respect to the tonotopic axis, and with respect to their electrophysiological properties [Soares and Carr, 2001; Soares et al., 2002] (fig. 5). NA contains four major morphological classes of neurons: planar, radiate, vertical, and stubby. Planar neurons are confined to an isofrequency band, whereas radiate neurons have dendrites that could extend across an isofrequency band. Vertical cells have long dendrites oriented perpendicularly to isofrequency bands. Stubby cells are confined to an isofrequency band because of their short dendrites. Representatives of all cell classes can be found throughout NA. Thus, a similar pattern of organization appears to have evolved in parallel in the cochlear nuclei of both birds and mammals, in which one population (planar, stubby, bushy) remains within an isofrequency band, another (radiate) extends across the isofrequency axis, and a third (vertical, marginal, octopus) has a dendritic orientation orthogonal to the isofrequency axis.

Although there is a similar organization in the ventral cochlear nucleus and NA, there are significant differences between the two structures. First, the majority of NA cells are stubby neurons that have no obvious counterpart within the multipolar cell types of the mammalian ventral cochlear nucleus. Second, there are many cell types within the mammalian ventral cochlear nucleus that are not included in Doucet and Ryugo's classification scheme, principally bushy cells, octopus cells, and small cell types [for review see Brawer and Morest, 1974; Cant and Morest, 1979; Rouiller and Ryugo, 1984; Cant, 1992; Rhode and Greenberg, 1992]. Bushy cells might be homologous to the cells of NM [Sullivan and Konishi, 1984], but we find no obvious morphological counterpart to octopus cells in NA. These neurons are located in the posterior part of the ventral cochlear nucleus and encode rapid transients [Oertel et al., 2000] (fig. 5). This is significant because both octopus cells and cells in NA respond to sound with onset responses [Sullivan, 1985; Warchol and Dallos, 1990]. Nevertheless, Golgi analyses of barn owl NA neurons and intracellular labeling of cells in chicken NA have not revealed cells with the characteristic octopus cell morphology – thick dendrites that extend across the incoming auditory nerve inputs [Oertel et al., 2000; Soares and Carr, 2001; Soares et al., 2002]. Thus, evolution of onset coding might not necessarily have produced identical solutions. Finally, avian cochlear nuclei do not show the small cell cap that characterizes mammals, just as they lack the cerebellar-like organization of the DCN. Only two small cells were found in NA [Soares and Carr,

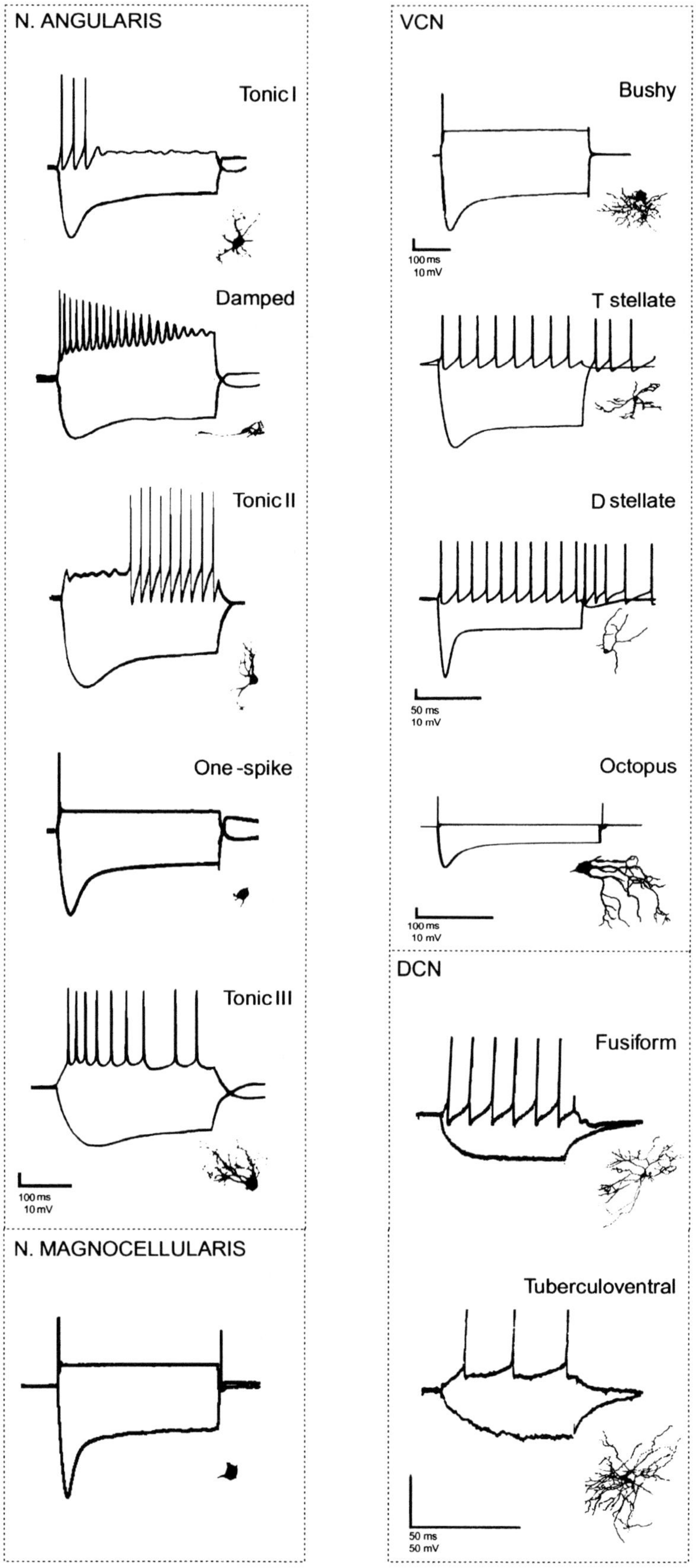

Fig. 5. Summary of current clamp responses in labeled cochlear nucleus neurons in chicken (**left**) and mammal (**right**). Avian NA data are taken from Soares et al. [2002], and NM data are from K. Macleod [unpublished data]. All avian scales: 100 ms and 10 mV. Currents injected: Tonic I, 150 and –500 pA; Damped, 150 and –500; Tonic II, 150 and –500; One spike, 200 and –500; Tonic III, 150 and –500; NM, 150 and –500. Mammalian information (right): Bushy cell: 100 ms, 10mV, 350 to –350 pA [physiology, Schwarz and Puil, 1997; anatomy, Cant, 1992]. T and D stellate: 50 ms and 10 mV, 100 to –600 pA [physiology, Fujino and Oertel, 2001; anatomy, T-stellate, Cant, 1992; anatomy, D-stellate, Doucet and Ryugo, 1997]. Octopus: 100 ms and 10 mV, 400 to –400 pA [physiology, Oertel et al., 2000; anatomy, Cant, 1992]. Tuberculoventral: 50 ms and 10 mV, 200 and –400 pA [physiology and anatomy, Zhang and Oertel, 1993]. Fusiform: 50 ms and 10 mV, 200 and –400 pA [physiology and anatomy, Zhang and Oertel, 1994]. Note that although responses from spherical bushy cells and NM neurons are similar, and both receive endbulb inputs from the auditory nerve, we have not placed them in adjacent rows because it is not clear whether they are homologous. Scales: 100 ms and 10 mV.

2001], and it appears that small cell types are not as various or numerous in NA as in the mammalian cochlear nucleus.

Heterogeneous Intracellular Physiological Responses

Neuronal responses to current injection in the cochlear nucleus show a heterogeneous repertoire of membrane characteristics in both birds and mammals (fig. 5). There is, however, no direct one-to-one correspondence between these traits, i.e., no single group of traits characterizes a cell type in both classes. The only exceptions are the features associated with temporal coding (see below). Because this suite of features characterizes temporal coding neurons in many auditory nuclei, including lemniscal nuclei [Wu, 1999], and is also found in temporal coding neurons in electric fish [Carr, 1986; Rashid et al., 2001], we cannot use their presence to argue for homology among cochlear nucleus neurons in birds and mammals.

In birds, stubby neurons respond with only one spike. Planar neurons show a damped response in the amplitude of their action potentials. Vertical neurons show a delayed response (tonic II) and radiate neurons show tonic responses to current injection (tonic I, III). The responses of stubby NA neurons are similar to those exhibited by both bushy and octopus cells in mammalian ventral cochlear nucleus [Wu and Oertel, 1984; Manis and Marx, 1991; Golding et al., 1999]. These cell types exhibit the depolarization-activated, dendrotoxin-sensitive, low-threshold K^+ conductance (LTC) that is activated at rest [see section 'Coincidence Detection ...', above, and Manis and Marx, 1991; Brew and Forsythe, 1995; Bal and Oertel, 2000]. A similar low threshold conductance could underlie the responses of NA one spike neurons, because it is also found in both NM and NL neurons [Reyes et al., 1994, 1996; Rathouz and Trussell, 1998] and in the irregularly firing principal cells of the tangential nucleus [Gamkrelidze et al., 1998, 2000]. These similarities suggest that stubby neurons, like bushy, octopus, NM and NL neurons, might mediate accurate transmission of temporal information [Oertel, 1999; Trussell, 1999].

There are no other NA neurons with direct one-to-one correspondence with mammalian cochlear nucleus neurons. The physiological characteristics that define planar neurons in birds are not seen in the cochlear nucleus neuron in mammals. A decrease in spike amplitude from the onset of depolarization is first seen in the mammalian inferior colliculus [Sivaramakrishnan and Oliver, 2001]. It is possible that similar computations could take place at different levels in the auditory system, or that different membrane properties might mediate the same physiological responses.

Radiate (tonic II) and pyramidal neurons in the mammalian DCN share two defining properties: fast inward rectification in response to hyperpolarizing current and delayed spike response to depolarizing current. Both produce subthreshold oscillations during depolarization pulses, similar to those in the pyramidal cells in the DCN [Manis and Molitor, 2001]. These low-frequency oscillations were blocked by tetrodotoxin in pyramidal cells (TTX, 500 n*M*). Voltage-gated Na^+ channels are therefore required to generate membrane oscillations, and Manis and Molitor suggest that they play a role in controlling spike timing in neurons when the membrane potential slowly approaches, or hovers near, spike threshold. Tonic II cells further resemble DCN pyramidal cells in that both exhibit similar delayed firing patterns in vitro [Kanold and Manis, 1999; Soares et al., 2002]. In vivo recordings will show whether these similar intracellular responses mediate similar responses to sound. Despite their similarities, tonic II and DCN pyramidal neurons have distinctly different projections [Young et al., 1988; Soares and Carr, 2001].

Radiate (tonic I and III) cells in NA share features with both D- and T-stellate neurons in the ventral cochlear nucleus. T- and D-stellates are physiologically differentiated in vitro by the shape of the action potential undershoot, which is rapidly repolarizing in D-stellates and scalloped in T-stellates (fig. 5). T- and D-stellates are further differentiated by differences in inward rectification, which is more prominent and more rapid in D-stellates [Fujino and Oertel, 2001]. Tonic I and III radiates are similarly distinguished. Tonic I neurons have prominent, rapid inward rectification, whereas Tonic III neurons have a scalloped undershoot and weak rectification. Despite these similarities, tonic cells in NA and neurons in the VCN differ in their transmitter phenotype and projections. They are unlikely to be homologous because T-stellates are excitatory neurons that project to the contralateral VCN, and D-stellate neurons are inhibitory glycinergic neurons that suppress activity in T-stellate cells [Wu and Oertel, 1984; Oertel et al., 1990; Saint Marie et al., 1991; Zhang and Oertel, 1993; Wickesberg et al., 1994; Gates et al., 1996; Moore et al., 1996; Ferragamo et al., 1998; Davis and Young, 2000]. In the barn owl, all NA neurons project to the midbrain [Soares and Carr, 2001] and none appear to be glycinergic. In these examples, there is again no direct one-to-one correspondence between the traits expressed in any cell type in birds and mammals.

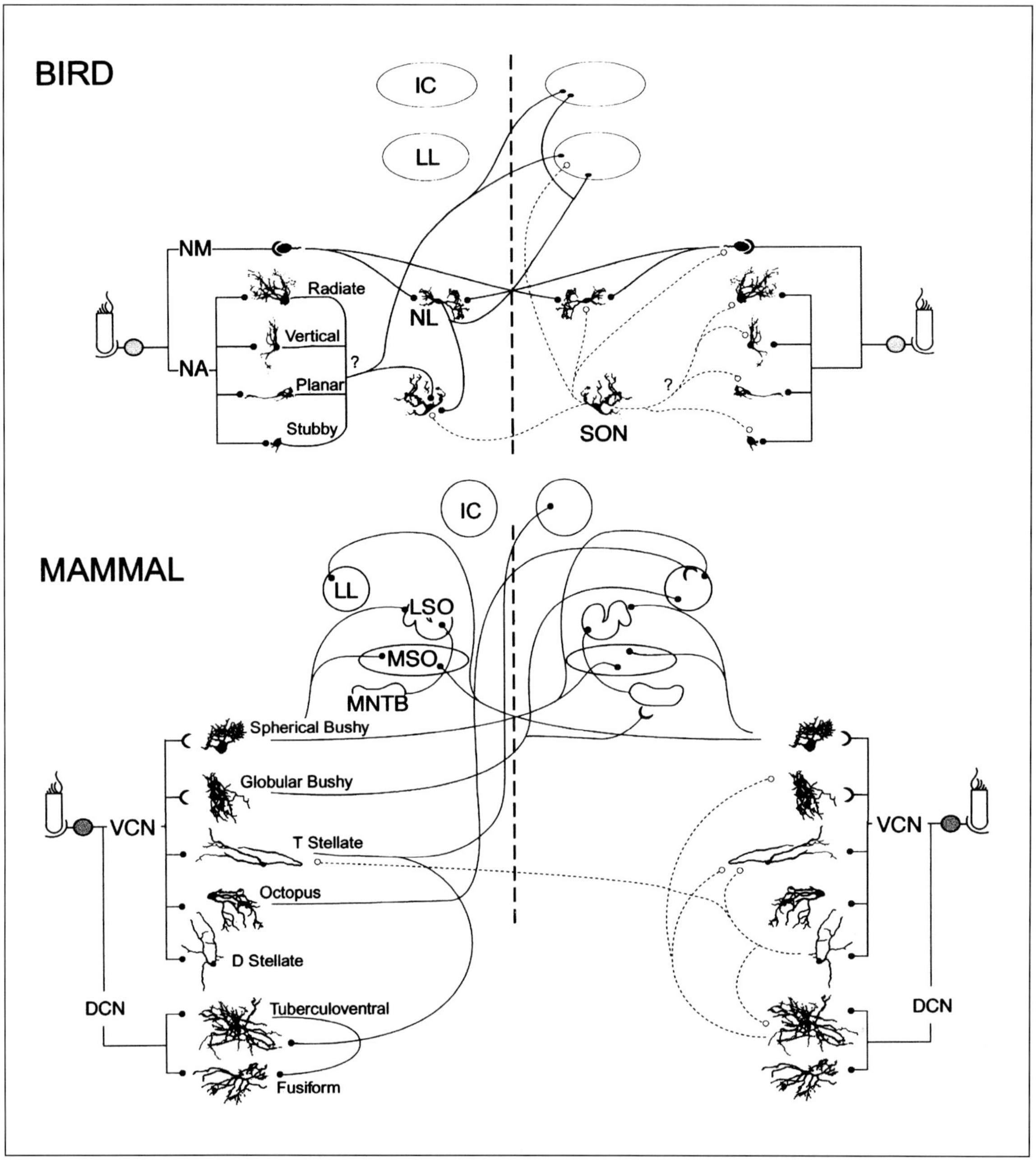

Fig. 6. Schematic summary of ascending connections and cell types of the avian and mammalian cochlear nuclei. Each connection displays a labeled cell body and shows the principal projections of each cell type. See text for citations. **Top:** Connections of the avian auditory system. NA (nucleus angularis) projections from Conlee and Parks [1986] and Yang et al. [1999]. **Bottom:** Connections of the mammalian cochlear nuclei. VCN (ventral cochlear nucleus) projections based on Oertel [1999], and DCN (dorsal cochlear nucleus) based on Wickesberg et al. [1991] and Zhang and Oertel [1994]. IC = Inferior colliculus; LL = lateral lemniscus; LSO = lateral superior olive; MNTB = medial nucleus of the trapezoid body; MSO = medial superior olive; NL = nucleus laminaris; NM = nucleus magnocellularis; SON = superior olivary nucleus. Note: It is not known whether the same SON cells project to both hindbrain and midbrain.

Ascending Lemniscal Projections

Avian and mammalian cochlear nuclei both share ascending lemniscal projections, but these differ in many respects (fig. 6). In birds and crocodilians, the NA projects to the contralateral posterior portion of the dorsal nucleus of the lateral lemniscus (LLDp, formerly VLVp) [Takahashi and Konishi, 1988a, b; Wild et al., 2001] and the inferior colliculus, while receiving GABAergic projections from the superior olivary nucleus (SON) [Conlee and Parks, 1986; Takahashi and Konishi 1988b; Yang et al., 1999]. The posterior division of the LLDp is the first site of binaural interactions in the intensity pathway of the barn owl and is where sensitivity to interaural level differences (ILDs) first appears [Moiseff and Konishi, 1983; Manley et al., 1988]. The pathways encoding ITDs and ILDs ultimately converge in the external nucleus of the inferior colliculus, where neurons are selective for combination of ITDs and ILDs [for review see Konishi, 2000]. In birds, the lemniscal projections of NA resemble octavolateralis projections in other vertebrates more than they resemble the lemniscal projections of the cells of the mammalian cochlear nucleus [for review see McCormick, 1992]. The projections of the mammalian cochlear nucleus are more variable [for reviews see Young et al., 1988; Cant, 1992; Romand and Avan, 1997; fig. 6].

At present it is not clear whether the features of the avian and mammalian cochlea nuclei described above mediate coding of some specific aspect of the auditory stimulus and are convergent, whether the similarities arose by chance, or whether they represent homologous features found in the common ancestor. Every neuron must be distinguished by a particular morphology and physiology, by embryological origins, location, and connections. Analysis of these characters in well-chosen species will be required to determine homology. It is not clear which characters should be given the most weight: projections can arborize and retract widely during development [Stanfield et al., 1982]; multiple transmitters can be expressed by any one neuron [Sato et al., 2000]; and patterns of channel expression appear to be driven by computational requirements. Furthermore, Turrigiano et al. [1994] have shown that there are activity-dependent changes in the intrinsic properties of cultured neurons, so neurons could be equipped with a suite of features suited for particular computations, and retain the ability to modify these over time [Desai et al., 1999].

Embryological studies are possible in chicken and quail, and recent studies have described the cell fate of chicken auditory brainstem neurons [Marin and Puelles, 1995; Cramer et al., 2000]. Similar descriptions of embryological origins are needed in other amniotes.

Conclusions

Birds and mammals share a common ancestor in the Carboniferous [Carroll, 1988]. Many amniote synapomorphies are widely interpreted as adaptations of these ancestors to life on land. We propose that some features we have described in the avian and mammalian auditory systems are apomorphic, or derived and different from the ancestral condition, and not homologous. The principal reason for arguing for the lack of homology is the separate development of true tympanic ears in their ancestors [Clack, 1997]. A second reason is that close comparisons of bird and mammal cochlear nuclei reveal many differences. A third is that the observed convergence of morphology and physiology of cochlear neurons is a plausible outcome of convergent evolution, because neurons in both birds and mammals experience similar constraints in detecting sound. Thus, although a common population of brainstem auditory neurons existed in the tetrapod ancestor, we propose that distinct evolutionary forces acted on these two groups allowing for the emergence of different ears and in turn, dissimilar organization in the brainstem. Some elements are likely to be homologous (such as large somatic terminals), but at present it is difficult to separate homology from convergence without careful analyses of all amniote groups.

Comparisons of temporal coding reveal shared computational principles. When compared with a simple integrate-and-fire unit, the auditory neurons that phase-lock, detect coincidences, and encode temporal patterns all exhibit a suite of physiological and morphological adaptations that suit them for their task. In fact, the essential features of auditory coding are very similar in birds and mammals (and probably in other animals as well). Comparative studies of temporal coding can therefore add to the discussion of whether neuronal function follows form. A case can be made for this in time coding neurons of the auditory brainstem of birds and mammals, and for phase coding neurons in weakly electric fish [Kawasaki, 2000].

If there are computational advantages to particular neuronal architectures, convergence should be expected. For example, we argue that the bitufted structure of coincidence detector neurons in birds and mammals is computationally advantageous. Therefore, morphological similarities might not support homology, but rather similar computational demands, and we can argue that the neurons of nucleus laminaris and MSO might have converged upon their similar form. In another example, it appears that large somatic terminals on NM or bushy cells are an ancestral feature of amniote auditory nerve. A shared pressure to encode higher frequency sounds might have driven the convergent appearance of complex endbulbs in archosaurs and mammals.

Finally, phenotypically different neurons can produce similar computations. Neurons can differ in the expression and/or distribution of their ionic channels and still behave similarly. Thus, there might be numerous acceptable ways to carry out a particular computation. These could be revealed by comparative studies.

Differences between classes, such as those documented in the avian and mammalian cochlear nuclei, might result from lack of homology, a long separation over evolutionary time, and/or different selective pressures. Differences within a group can reveal how neural circuits evolve. The nervous system is evolutionarily conservative and small changes in structure can lead to profound changes in function and behavior [Nishikawa, 1997]. Within birds, there are differences in the subcellular distribution of potassium channels in chick and owl delay lines. The changes in presynaptic localization might be a specialization for enabling neurons in owl NM to transmit high frequency temporal information with little jitter [Parameshwaran et al., 2001]. Nishikawa [1997] has proposed that these types of interspecific differences could occur as small changes superimposed upon a basic developmental plan. Understanding how this type of change take place in the context of increased behavioral function should illuminate mechanisms by which neural circuits evolve.

Acknowledgments

We gratefully acknowledge helpful discussions with Drs. J. Simon, H. Eisthen, and K. Nishikawa. This work was supported by NIH DCD00436 to C.E.C.

References

Agmon-Snir, H., C.E. Carr, and J. Rinzel (1998) The role of dendrites in auditory coincidence detection. Nature, *393:* 268–272.

Amagai, S., M.A. Friedman, and C.D. Hopkins (1998) Time-coding in the midbrain of mormyrid electric fish I: Physiology and anatomy of cells in the nucleus exterolateralis pars anterior. J. Comp. Physiol. A, *182:* 115–130.

Bal, R., and D. Oertel (2000) Hyperpolarization-activated, mixed-cation current (I_h) in octopus cells of the mammalian cochlear nucleus. J. Neurophysiol., *84:* 806–817.

Batra, R., S. Kuwada, and D.C. Fitzpatrick (1997) Sensitivity to interaural temporal disparities of low- and high-frequency neurons in the superior olivary complex. I. Heterogeneity of responses. J. Neurophysiol., *78:* 1222–1236.

Bell, C.C., and T. Szabo (1986) Electroreception in mormyrid fish: Central anatomy. *In* Electroreception (ed. by T.H. Bullock and W. Heiligenberg), Wiley, New York, pp. 375–421.

Boord, R.L. (1969) The anatomy of the avian auditory system. Ann. N.Y. Acad. Sci., *167:* 147–155.

Borst, J.G., and B. Sakmann (1996) Calcium influx and transmitter release in a fast CNS synapse. Nature, *383:* 431–434.

Borst, J.G., F. Helmchen, and B. Sakmann (1995) Pre- and postsynaptic whole-cell recordings in the medial nucleus of the trapezoid body of the rat. J. Physiol., *489:* 825–840.

Brand, A., O. Behrend, T. Marquardt, D. McAlpine, and B. Grothe (2002) Precise inhibition is essential for microsecond interaural time difference coding. Nature, *417:* 543–547.

Brawer, J.R., and D.K. Morest (1974) Relations between auditory nerve endings and cell types in the cat's anteroventral cochlear nucleus seen with Golgi method and Nomarski optics. J. Comp. Neurol., *160:* 491–506.

Brenowitz, S., and L.O. Trussell (2001) Minimizing synaptic depression by control of release probability. J. Neurosci., *21:* 1857–1867.

Brew, H.M., and I.D. Forsythe (1995) Two voltage-dependent K^+ conductances with complementary functions in postsynaptic integration at a central auditory synapse. J. Neurosci., *15:* 8011–8022.

Cant, N.B. (1992) The cochlear nucleus: Neuronal types and their synaptic organization. *In* The Mammalian Auditory Pathway: Neuroanatomy (ed. by D.B. Webster, A.N. Popper and R.R. Fay), Springer-Verlag, New York, pp. 66–116.

Cant, N.B., and R.L. Hyson (1992) Projections from the lateral nucleus of the trapezoid body to the medial superior olivary nucleus in the gerbil. Hear. Res., *58:* 26–34.

Cant, N.B., and D.K. Morest (1979) Organization of the neurons in the anterior division of the anteroventral cochlear nucleus of the cat. Light-microscopic observations. Neuroscience, *4:* 1909–1923.

Carr, C.E. (1986) Time coding in electric fish and barn owls. Brain Behav. Evol., *28:* 122–133.

Carr, C.E., and S. Amagai (1996) Processing of temporal information in the brain. *In* Time, Internal Clocks and Movement (ed. by M.A. Pastor and J. Artieda), Elsevier, New York, pp. 27–52.

Carr, C.E., and R.E. Boudreau (1993) Organization of the nucleus magnocellularis and the nucleus laminaris in the barn owl: Encoding and measuring interaural time differences. J. Comp. Neurol., *316:* 223–243.

Carr, C.E., and M. Konishi (1990) A circuit for detection of interaural time differences in the brainstem of the barn owl. J. Neurosci., *10:* 3227–3246.

Carr, C.E., M.F. Kubke, D.P. Massoglia, S.M. Cheng, L.L. Rigby, and A. Moiseff (1997) Development of temporal coding circuits in the barn owl. *In* Psychophysical and Physiological Advances in Hearing (ed. by A. R. Palmer, A. Rees, Q. Summerfield and R. Meddis), Whurr Publishers, London, pp. 344–351.

Carroll, R.L. (1988) Vertebrate Paleontology and Evolution. W.H. Freeman, New York.

Clack, J.A. (1997) The evolution of tetrapod ears and the fossil record. Brain Behav. Evol., *50:* 198–212.

Colburn, H.S., Y. Han, and C.P. Culotta (1990) Coincidence model of MSO responses. Hear. Res., *49:* 335–355.

Conlee, J.W., and T.N. Parks (1986) Origin of ascending projections to the nucleus mesencephalicus lateralis pars dorsalis in the chicken. J. Comp. Neurol., *367:* 96–113.

Cramer, K.S., S.E. Fraser and E.W. Rubel (2000) Embryonic origins of auditory brain-stem nuclei in the chick hindbrain. Dev. Biol., *224:* 138–151.

Davis, K.A., and E.D. Young (2000) Pharmacological evidence of inhibitory and disinhibitory neuronal circuits in dorsal cochlear nucleus. J. Neurophysiol., *83:* 926–940.

Desai, N.S., L.C. Rutherford, and G.G. Turrigiano (1999) Plasticity in the intrinsic excitability of cortical pyramidal neurons. Nat. Neurosci., *3:* 515–520.

Doucet, J.R., and D.K. Ryugo (1997) Projections from the ventral cochlear nucleus to the dorsal cochlear nucleus in rats. J. Comp. Neurol., *385:* 245–264.

Ferragamo, M.J., N.L. Golding, and D. Oertel (1998) Synaptic inputs to stellate cells in the ventral cochlear nucleus. J. Neurophysiol., *79:* 51–63.

Fujino, H., and D. Oertel (2001) Cholinergic modulation of stellate cells in the mammalian ventral cochlear nucleus. J. Neurosci., *21:* 7372–7383.

Funabiki, K., K. Koyano, and K.H. Ohmori (1998) The role of GABAergic inputs for coincidence detection in the neurones of nucleus laminaris of the chick. J. Physiol., *508:* 851–869.

Gamkrelidze, G., C. Giaume, and K.D. Peusner (1998) The differential expression of low-threshold sustained potassium current contributes to the distinct firing patterns in embryonic central vestibular neurons. J. Neurosci., *18:* 1449–1464.

Gamkrelidze, G., C. Giaume, and K.D. Peusner (2000) Firing properties and dendrotoxin-sensitive sustained potassium current in vestibular nuclei neurons of the hatchling chick. Exp. Brain Res., *134:* 398–401.

Gardner, S., L.O. Trussell, and D. Oertel (1999) Time course and permeation of synaptic AMPA receptors in cochlear nucleus neurons correlate with input. J. Neurosci., *19:* 8721–8729.

Gardner, S., L.O. Trussell, and D. Oertel (2001) Correlation of AMPA receptor subunit composition with synaptic input in the mammalian cochlear nuclei. J. Neurosci., *21:* 7428–7437.

Gates, T.S., D.L. Weedman, T. Pongstaporn, and D.K. Ryugo (1996) Immunocytochemical localization of glycine in a subset of cartwheel cells of the dorsal cochlear nucleus in rats. Hear. Res., *96:* 157–166.

Gerstner, W., R. Kempter, J.L. van Hemmen, and H. Wagner (1996) A neuronal learning rule for sub-millisecond temporal coding. Nature, *383:* 76–81.

Goldberg, J.M., and P.B. Brown (1969) Response of binaural neurons of dog superior olivary complex to dichotic tonal stimuli: Some physiological mechanisms of sound localization. J. Neurophysiol., *32:* 613–636.

Golding, N.L., D. Robertson, and D. Oertel (1995) Recordings from slices indicate that octopus cells of the cochlear nucleus detect coincident firing of auditory nerve fibers with temporal precision. J. Neurosci., *15:* 3138–3153.

Golding, N.L., M.J. Ferragamo, and D. Oertel (1999) Role of intrinsic conductances underlying responses to transients in octopus cells of the cochlear nucleus. J. Neurosci., *19:* 2897–2905.

Grigg, J.J., H.M. Brew, and B.L. Tempel (2000) Differential expression of voltage-gated potassium channel genes in auditory nuclei of the mouse brainstem. Hear. Res., *140:* 77–90.

Grothe, B., and D.H. Sanes (1994) Synaptic inhibition influences the temporal coding properties of medial superior olivary neurons: An in vitro study. J. Neurosci., *14:* 1701–1709.

Hafter, E.R., and C. Trahiotis (1997) Functions of the binaural system. *In* Handbook of Acoustics (ed. by M. Crocker), Wiley, New York, pp. 297–345.

Heffner, R.S. (1997) Comparative study of sound localization and its anatomical correlates in mammals. Acta Otolaryngol. Suppl., *532:* 46–53.

Heffner, R.S., and H.E. Heffner (1992) Evolution of sound localization in mammals. *In* The Evolutionary Biology of Hearing (ed. by D.B. Webster, R. R. Fay and A.N. Popper), Springer-Verlag, New York, pp. 691–716.

Heiligenberg, W. (1989) Coding and processing of electrosensory information in gymnotiform fish. J. Exp. Biol., *146:* 255–275.

Hill, K.G., G. Stange, and J. Mo (1989) Temporal synchronization in the primary auditory response in the pigeon. Hear. Res., *39:* 63–74.

Isaacson, J.S., and B. Walmsley (1996) Amplitude and time course of spontaneous and evoked excitatory postsynaptic currents in bushy cells of the anteroventral cochlear nucleus. J. Neurophysiol., *76:* 1566–1571.

Jeffress, L.A. (1948) A place theory of sound localization. J. Comp. Physiol. Psych., *41:* 35–39.

Jhaveri, S., and D.K. Morest (1982) Neuronal architecture in nucleus magnocellularis of the chicken auditory system with observations on nucleus laminaris: A light and electron microscope study. Neuroscience, *7:* 809–836.

Joris, P.X., P.H. Smith, and T.C.T. Yin (1998) Coincidence detection in the auditory system: 50 years after Jeffress. Neuron, *21:* 1235–1238.

Kanold, P.O., and P.B. Manis (1999) Transient potassium currents regulate the discharge patterns of dorsal cochlear nucleus pyramidal cells. J. Neurosci., *19:* 2195–2208.

Kapfer, C., A.H. Seidl, H. Schweizer, and B. Grothe (2002) Experience-dependent refinement of inhibitory inputs to auditory coincidence-detector neurons. Nat. Neurosci., *5:* 247–253.

Kawasaki, M. (2000) Phylogenetic evolution of computational algorithms: Nonparametric approach to knowledge. Discovery, *8:* 77–80.

Kawasaki, M., and Y.X. Guo (1996) Neuronal circuitry for comparison of timing in the electrosensory lateral line lobe of the African wave-type electric fish *Gymnarchus niloticus*. J. Neurosci., *16:* 380–391.

Kawasaki, M., G. Rose, and W. Heiligenberg (1988) Temporal hyperacuity in single neurons of electric fish. Nature, *336:* 173–176.

Konishi, M. (1973) How the owl tracks its prey. Am. Sci., *61:* 414–424.

Konishi, M. (1991) Deciphering the brain's codes. Neural Comput., *3:* 1–18.

Konishi, M. (2000) Study of sound localization by owls and its relevance to humans. Comp. Biochem. Physiol., *126:* 459–469.

Köppl, C. (1997) Phase locking to high frequencies in the auditory nerve and cochlear nucleus magnocellularis of the barn owl, *Tyto alba*. J. Neurosci., *17:* 3312–3321.

Köppl, C., C.E. Carr, and D. Soares (2001) Diversity of response patterns in the cochlear nucleus angularis (NA) of the barn owl. Assoc. Res. Otolaryngol. Abstr., *21:* 709.

Kuwabara, N., and J.M. Zook (1992) Projections to the medial superior olive from the medial and lateral nuclei of the trapezoid body in rodents and bats. J. Comp. Neurol., *324:* 522–538.

Li, W., L.K. Kaczmarek, and T.M. Perney (2001) Localization of two high-threshold potassium channel subunits in the rat central auditory system. J. Comp. Neurol., *437:* 196–218.

Mainen, Z.F., and T.J. Sejnowski (1995) Reliability of spike timing in neocortical neurons. Science, *268:* 1503–1506.

Mainen, Z.F., and T.J. Sejnowski (1996) Influence of dendritic structure on firing pattern in model neocortical neurons. Nature, *382:* 363–366.

Manis, P.B., and S.O. Marx (1991) Outward currents in isolated ventral cochlear nucleus neurons. J. Neurosci., *11:* 2865–2880.

Manis, P.B., and S.C. Molitor (2001) Subthreshold TTX-sensitive voltage oscillations in DCN pyramidal cells. Assoc. Res. Otolaryngol. Abstr., *21:* 599.

Manley, G.A., C. Köppl and M. Konishi (1988) A neural map of interaural intensity difference in the brainstem of the barn owl. J Neurosci., *8:* 2665–2677.

Marin, F., and L. Puelles (1995) Morphological fate of rhombomeres in quail/chick chimeras: A segmental analysis of hindbrain nuclei. Eur. J. Neurosci., *7:* 1714–1738.

McCormick, C. (1992) Evolution of central auditory pathways in anamniotes. *In* The Evolutionary Biology of Hearing (ed. by D.B. Webster, A.N. Popper and R.R. Fay), Springer-Verlag, New York, pp. 323–335.

Moiseff, A., and M. Konishi (1983) Binaural characteristics of units in the owl's brainstem auditory pathway: Precursors of restricted spatial receptive fields. J. Neurosci., *12:* 2553–2562.

Monsivais, P., L. Yang, and E.W. Rubel (2000) GABAergic inhibition in nucleus magnocellularis: Implications for phase locking in the avian auditory brainstem. J. Neurosci., *20:* 2954–2963.

Moore, J.K., K.K. Osen, J. Storm-Mathisen, and O.P. Ottersen (1996) γ-Aminobutyric acid and glycine in the baboon cochlear nuclei: An immunocytochemical colocalization study with reference to interspecies differences in inhibitory systems. J. Comp. Neurol., *369:* 497–519.

Mugnaini, E., A.S. Berrebi, A.L. Dahl, and J.I. Morgan (1987) The polypeptide PEP-19 is a marker for Purkinje neurons in cerebellar cortex and cartwheel neurons in the dorsal cochlear nucleus. Arch. Ital. Biol., *126:* 41–67.

Mugnaini, E., M.R. Dino, and D. Jaarsma (1997) The unipolar brush cells of the mammalian cerebellum and cochlear nucleus: Cytology and microcircuitry. Prog. Brain Res., *114:* 131–150.

Nishikawa, K.C. (1997) Emergence of novel functions during brain evolution. Bioscience, *47:* 341–354.
Oertel, D. (1999) The role of timing in the brain stem auditory nuclei of vertebrates. Annu. Rev. Physiol., *61:* 497–519.
Oertel, D., R. Bal, S. Gardner, P. Smith, and P. Joris (2000) Detection of synchrony in the activity of auditory nerve fibers by octopus cells of the mammalian cochlear nucleus. Proc. Nat. Acad. Sci. USA, *97:* 11773–11779.
Oertel, D., S.H. Wu, M.W. Garb, and C. Dizack (1990) Morphology and physiology of cells in slice preparations of the posteroventral cochlear nucleus of mice. J. Comp. Neurol., *295:* 136–154.
Overholt, E.M., E.W. Rubel, and R.L. Hyson (1992) A circuit for coding interaural time differences in the chick brain stem. J. Neurosci., *12:* 1698–1708.
Parameshwaran, S., C.E. Carr, and T.M. Perney (2001) Expression of the Kv3.1 potassium channel in the avian auditory brainstem. J. Neurosci., *21:* 485–494.
Parks, T.N. (2000) The AMPA receptors of auditory neurons. Hear. Res., *147:* 77–91.
Peña, J.L., S. Viete, Y. Albeck, and M. Konishi (1996) Tolerance to sound intensity of binaural coincidence detection in the nucleus laminaris of the owl. J. Neurosci., *16:* 7046–7054.
Perney, T.M., and L.K. Kaczmarek (1997) Localization of a high threshold potassium channel in the rat cochlear nucleus. J. Comp. Neurol., *386:* 178–202.
Raman, I.M., and L.O. Trussell (1992) The kinetics of the response to glutamate and kainate in neurons of the avian cochlear nucleus. Neuron, *9:* 173–186.
Rashid, A.J., E. Morales, R.W. Turner, and R. J. Dunn (2001) The contribution of dendritic Kv3 K^+ channels to burst threshold in a sensory neuron. J. Neurosci., *21:* 125–135.
Rathouz, M., and L.O. Trussell (1998) Characterization of outward currents in neurons of the avian nucleus magnocellularis. J. Neurophysiol., *80:* 2824–2835.
Ravindranathan, A., T.N. Parks, and M.S. Rao (1996) Flip and flop isoforms of chick brain AMPA receptor subunits: Cloning and analysis of expression patterns. NeuroReport, *7:* 2707–2711.
Reyes, A.D., E.W. Rubel, and W.J. Spain (1994) Membrane properties underlying the firing of neurons in the avian cochlear nucleus. J. Neurosci., *14:* 5352–5364.
Reyes, A.D., E.W. Rubel, and W.J. Spain (1996) In vitro analysis of optimal stimuli for phase-locking and time-delayed modulation of firing in avian nucleus laminaris neurons. J. Neurosci., *16:* 993–1007.
Rhode, W.S., and S. Greenberg (1992) Physiology of the cochlear nuclei. *In* The Mammalian Auditory Pathway: Neurophysiology (ed. by A.N. Popper and R.R. Fay), Springer-Verlag, Heidelberg, pp. 53–120.
Romand, R., and P. Avan (1997) Anatomical and functional aspects of the cochlear nucleus. *In* The Central Auditory System (ed. by G. Ehret and R. Romand), Oxford University Press, Oxford, UK, pp. 97–192.
Rouiller, E., and D. Ryugo (1984) Intracellular marking of physiologically characterized cells in the ventral cochlear nucleus of the cat. J. Comp. Neurol., *225:* 167–186.
Ryugo, D.K. (1991) The auditory nerve: Peripheral innervation, cell body morphology, and central projections. *In* The Mammalian Auditory Pathway: Neuroanatomy (ed. by D.B. Webster, A.N. Popper and R.R. Fay), Springer, New York, pp. 23–65.
Ryugo, D.K., M.M. Wu, and T. Pongstaporn (1996) Activity-related features of synapse morphology: A study of endbulbs of Held. J. Comp. Neurol., *365:* 141–158.
Sabatini, B.L., and W.G. Regehr (1999) Timing of synaptic transmission. Ann. Rev. Physiol., *61:* 521–542.
Sachs, M.B., and J.M. Sinnott (1978) Responses to tones of single cells in nucleus magnocellularis and nucleus angularis of the redwing blackbird *(Agelaius phoeniceus)*. J. Comp. Physiol., *126:* 347–361.
Sachs, M.B., N.K. Woolf, and J.M. Sinnott (1980) Response properties of neurons in the avian auditory system: Comparisons with mammalian homologues and consideration of the neural encoding of complex stimuli. *In* Comparative Studies of Hearing in Vertebrates (ed. by A.N. Popper and R.R. Fay), Springer-Verlag, New York, pp. 323–353.
Saint Marie, R.L., C.G. Benson, E.M. Ostapoff, and D.K. Morest (1991) Glycine immunoreactive projections from the dorsal to the anteroventral cochlear nucleus. Hear. Res., *51:* 11–28.
Sato, K., S. Shiraishi, H. Nakagawa, H. Kuriyama, and R.A. Altschuler (2000) Diversity and plasticity in amino acid receptor subunits in the rat auditory brain stem. Hear. Res., *147:* 137–144.
Schwarz, D.W.F., and E. Puil (1997) Firing properties of spherical bushy cells in the AVCN of the gerbil. Hear. Res., *114:* 127–138.
Simon, J.Z., C.E. Carr, and S.A. Shamma (1999) A dendritic model of coincidence detection in the avian brainstem. Neurocomputing, *26–27:* 263–269.
Simon, J.Z., C.E. Carr, and S.A. Shamma (2000) Biophysical model of coincidence detection in single Nucleus laminaris neurons. Assoc. Res. Otolaryngol. Abstr., *20:* 572.
Sivaramakrishnan, S., and D.L. Oliver (2001) Distinct K^+ currents result in physiologically distinct cell types in the inferior colliculus of the rat. J. Neurosci., *21:* 2861–2877.
Smith, D.J., and E.W. Rubel (1979) Organization and development of brain stem auditory nuclei of the chicken: Dendritic gradients in nucleus laminaris. J. Comp. Neurol., *186:* 213–240.
Smith, P.H. (1995) Structural and functional differences distinguish principal from nonprincipal cells in the guinea pig MSO slice. J. Neurophysiol., *73:* 1653–1667.
Soares, D., and C.E. Carr (2001) The cytoarchitecture of the nucleus angularis of the barn owl *(Tyto alba)*. J. Comp. Neurol., *429:* 192–205.
Soares, D., R. Chitwood, R.L. Hyson, and C.E. Carr (2002) The intrinsic physiological and morphological characteristics of the chicken nucleus angularis. J. Neurophysiol., in press.
Stanfield, B.B., D.D. O'Leary, and C. Fricks (1982) Selective collateral elimination in early postnatal development restricts cortical distribution of rat pyramidal tract neurones. Nature, *298:* 371–373.
Sullivan, W.E. (1985) Classification of response patterns in cochlear nucleus of barn owl: Correlation with functional response properties. J. Neurophysiol., *53:* 201–216.
Sullivan, W.E., and M. Konishi (1984) Segregation of stimulus phase and intensity coding in the cochlear nucleus of the barn owl. J. Neurosci., *4:* 1787–1799.
Szpir, M.R., S. Sento, and D.K. Ryugo (1990) The central projections of the cochlear nerve fibers in the alligator lizard. J. Comp. Neurol., *295:* 530–547.
Takahashi, T., and M. Konishi (1988a) The projections of nucleus angularis and nucleus laminaris to the lateral lemniscal nuclear complex of the barn owl. J. Comp. Neurol., *274:* 212–238.
Takahashi, T.T., and M. Konishi (1988b) Projections of the cochlear nuclei and nucleus laminaris to the inferior colliculus of the barn owl. J. Comp. Neurol., *274:* 190–211.
Takahashi, T., A. Moiseff, and M. Konishi (1984) Time and intensity cues are processed independently in the auditory system of the owl. J. Neurosci., *4:* 1781–1786.
Taschenberger, H., and H. von Gersdorff (2000) Fine-tuning an auditory synapse for speed and fidelity: Developmental changes in presynaptic waveform, EPSC kinetics, and synaptic plasticity. J. Neurosci., *20:* 9162–9173.
Trussell, L.O. (1997) Cellular mechanisms for preservation of timing in central auditory pathways. Curr. Opin. Neurobiol., *7:* 487–492.
Trussell, L.O. (1999) Synaptic mechanisms for coding timing in auditory neurons. Ann. Rev. Physiol., *61:* 477–496.
Turecek, R., and L.O. Trussell (2000) Control of synaptic depression by glutamate transporters. J. Neurosci., *20:* 2054–2063.
Turrigiano, G., L.F. Abbott, and E. Marder (1994) Activity-dependent changes in the intrinsic properties of cultured neurons. Science, *264:* 974–977.
Wang, L.Y., L. Gan, I.D. Forsythe, and L.K. Kaczmarek (1998) Contribution of the Kv3.1 potassium channel to high-frequency firing in mouse auditory neurones. J. Physiol., *509:* 183–194.
Warchol, M.E., and P. Dallos (1990) Neural coding in the chick cochlear nucleus. J. Comp. Physiol., *166:* 721–734.
Wickesberg, R.E., D. Whitlon, and D. Oertel (1991) Tuberculoventral neurons project to the multipolar cell area but not to the octopus cell area of the posteroventral cochlear nucleus. J. Comp. Neurol., *313:* 457–468.

Wickesberg, R.E., D. Whitlon, and D. Oertel (1994) In vitro modulation of somatic glycine-like immunoreactivity in presumed glycinergic neurons. J. Comp. Neurol., *339:* 311–327.

Wilczynski, W. (1984) Central neural systems subserving a homoplasous periphery. Am. Zool., *24:* 755–763.

Wild, J.M., M.F. Kubke, and C.E. Carr (2001) Tonotopic and somatotopic representation in the nucleus basalis of the barn owl, *Tyto alba.* Brain Behav. Evol., *57:* 39–62.

Wright, D.D., and D.K. Ryugo (1996) Mossy fiber projections from the cuneate nucleus to the cochlear nucleus in the rat. J. Comp. Neurol., *365:* 159–172.

Wu, S.H. (1999) Physiological properties of neurons in the ventral nucleus of the lateral lemniscus of the rat: Intrinsic membrane properties and synaptic responses. J. Neurophysiol., *81:* 2872–2884.

Wu, S.H., and D. Oertel (1984) Intracellular injection with horseradish peroxidase of physiologically characterized stellate and bushy cells in slices of mouse anteroventral cochlear nucleus. J. Neurosci., *4:* 1577–1588.

Yang, L., P. Monsivais, and E.W. Rubel (1999) The superior olivary nucleus and its influence on nucleus laminaris: A source of inhibitory feedback for coincidence detection in the avian auditory brainstem. J. Neurosci., *19:* 2313–2325.

Yin, T.C.T., and J.C.K. Chan (1990) Interaural time sensitivity in medial superior olive of cat. J. Neurophysiol., *64:* 465–488.

Young, S.R., and E.W. Rubel (1983) Frequency specific projections of individual neurons in chick brain stem auditory nuclei. J. Neurosci., *7:* 1373–1378.

Young, E.D., W.P. Shofner, J.A. White, J.M. Robert, and H.F. Voight (1988) Response properties of cochlear nucleus neurons in relationship to physiological mechanisms. *In* Auditory Function: Neurobiological Bases of Hearing (ed. by G.M. Edelman, W.E. Gall and W.M. Cowan), Wiley, New York, pp. 277–312.

Zhang, S., and D. Oertel (1993) Tuberculoventral cells of the dorsal cochlear nucleus of mice: Intracellular recordings in slices. J. Neurophysiol., *69:* 1409–1421.

Zhang, S., and D. Oertel (1994) Neuronal circuits associated with the output of the dorsal cochlear nucleus through fusiform cells. J. Neurophysiol., *71:* 914–930.

Zhang, S., and L.O. Trussell (1994) A characterization of excitatory postsynaptic potentials in the avian nucleus magnocellularis. J. Neurophysiol., *72:* 705–718.

Brain, Behavior and Evolution

Brain Behav Evol 2002;59:312–326

Evolution of Cerebellum-Like Structures

Curtis C. Bell

Neurological Sciences Institute, Oregon Health and Sciences University, Beaverton, Oreg., USA

Key Words

Cerebellum · Cerebellum-like structures · Evolution · Homology · Convergence · Syngeny

Abstract

All vertebrate brains have a cerebellum, and most of them have one or more additional structures that are histologically similar to the cerebellum. The cerebellum-like structures include the medial octavolateral nucleus in most aquatic vertebrates; the dorsal octavolateral nucleus in many aquatic vertebrates with an electrosensory system; the marginal layer of the optic tectum in ray-finned fishes; electrosensory lobes in the few groups of advanced bony fish with an electrosensory system; the rostrolateral nucleus of the thalamus in a few widely scattered groups of bony fish; and the dorsal cochlear nucleus in all mammals except monotremes. All of these structures receive topographically organized sensory input in their deep layers. Purkinje-like cells receive the sensory input near their cell bodies. These cells extend apical dendrites up into the molecular layer where they receive synaptic input from parallel fibers. The cerebellum itself can be included within this characterization by considering the climbing fiber as at least in part a conveyor of sensory information and by recalling that climbing fibers in more basal vertebrates terminate on smooth dendrites close to the soma. Physiological findings from three different systems suggest the hypothesis that cerebellum-like structures remove predictable features from the sensory inflow. Phylogenetic homology can explain the similarities across different taxa for some types of cerebellum-like structures, but similarities within other types cannot be explained in this way. Moreover, phylogenetic homology cannot explain the similarities among different types of cerebellum-like structures. Evolutionary convergence provides the best explanation for all these similarities that cannot be explained by homology. The convergence is almost surely constrained by the availability of a genetic-developmental program for creating cerebellum-like circuitry and by the need within many different systems for the type of information processing that cerebellum-like circuitry can provide.

Introduction

The cerebellum is present in all vertebrate brains [Nieuwenhuys, 1967; Larsell, 1967], and most vertebrate brains also have additional structures that are histologically similar to the cerebellum. The cerebellum and cerebellum-like structures vary widely among the different vertebrate taxa and neural systems, but their basic architecture is distinct and easily recognized. The distinctiveness of the architecture and the similarity across so many different taxa and systems lead naturally to an inquiry into the evolutionary and developmental origins of the similarities, and make cerebellum-like structures an attractive subject for such an inquiry. The most basic issue is whether the similarities among these structures are best explained by homology, that is by evolution from a com-

KARGER

Fax + 41 61 306 12 34
E-Mail karger@karger.ch
www.karger.com

Accessible online at:
www.karger.com/journals/bbe

Curtis C. Bell
Neurological Sciences Institute
Oregon Health and Sciences Institute
505 N.W. 185th Ave., Beaverton, OR 97006 (USA)
Tel. +1 503 418 2575, Fax +1 503 418 2501, E-Mail bellc@ohsu.edu

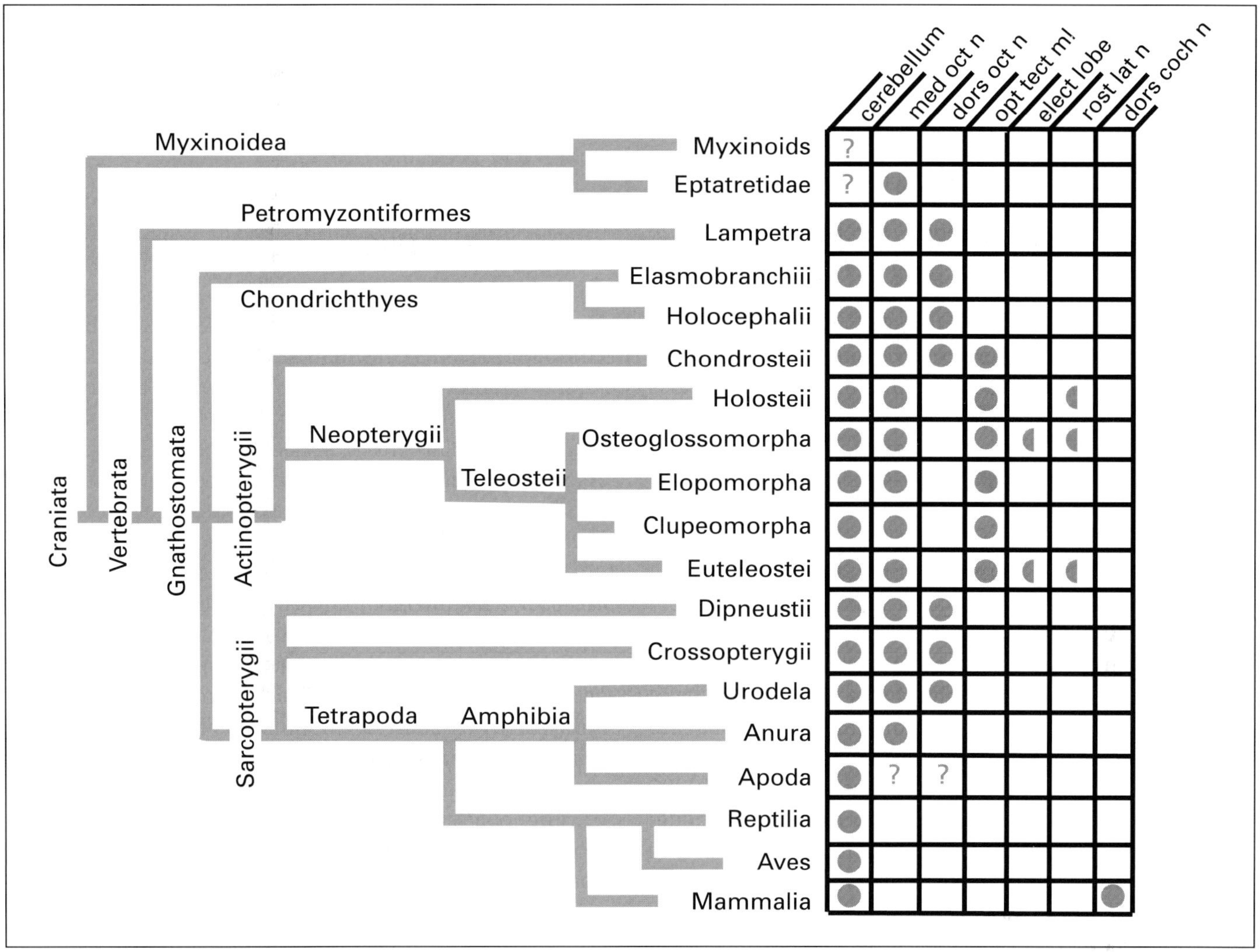

Fig. 1. Occurrences of cerebellum-like structures in different vertebrate groups. A filled circle means the structure is present in all or almost all the members of that group. A filled half circle means the structure is present only sporadically in that group. A question mark means that presence of the structure in that group is either unclear or controversial. Med oct n = medial octavolateral nucleus; dors oct n = dorsal octavolateral nucleus; opt tect ml = optic tectum marginal layer; elect lobe = electrosensory lobe; rost lat n = rostrolateral nucleus of thalamus; dors coch n = dorsal cochlear nucleus.

mon ancestral cerebellum-like structure, or are best explained by convergent evolution as constrained perhaps by available genetic mechanisms and by the need for similar types of information processing.

This review describes the different cerebellum-like structures of craniates and explores the issue of their evolutionary origin. The category 'craniates' includes the vertebrates and the myxinoids (hagfish) (see fig. 1). The myxinoids were once classified as vertebrates, but are now considered to be a sister group of the vertebrates [Liem et al., 2001]. The new classification scheme replaces an earlier one in which the class Vertebrata was divided into two groups, the Agnatha ('jawless' animals) and the Gnathostomata ('jawed' animals). The 'Agnatha' included the myxinoids and petromyzontids (lampreys). The 'Gnathostomata' included all other vertebrates. The new classification scheme reflects current phylogenetic thinking which places the petromyzontids and gnathostomes together as vertebrates and considers the myxinoids as a separate group.

The cerebellum-like structures included in this review are as follows: the *medial octavolateral nucleus* (MON),

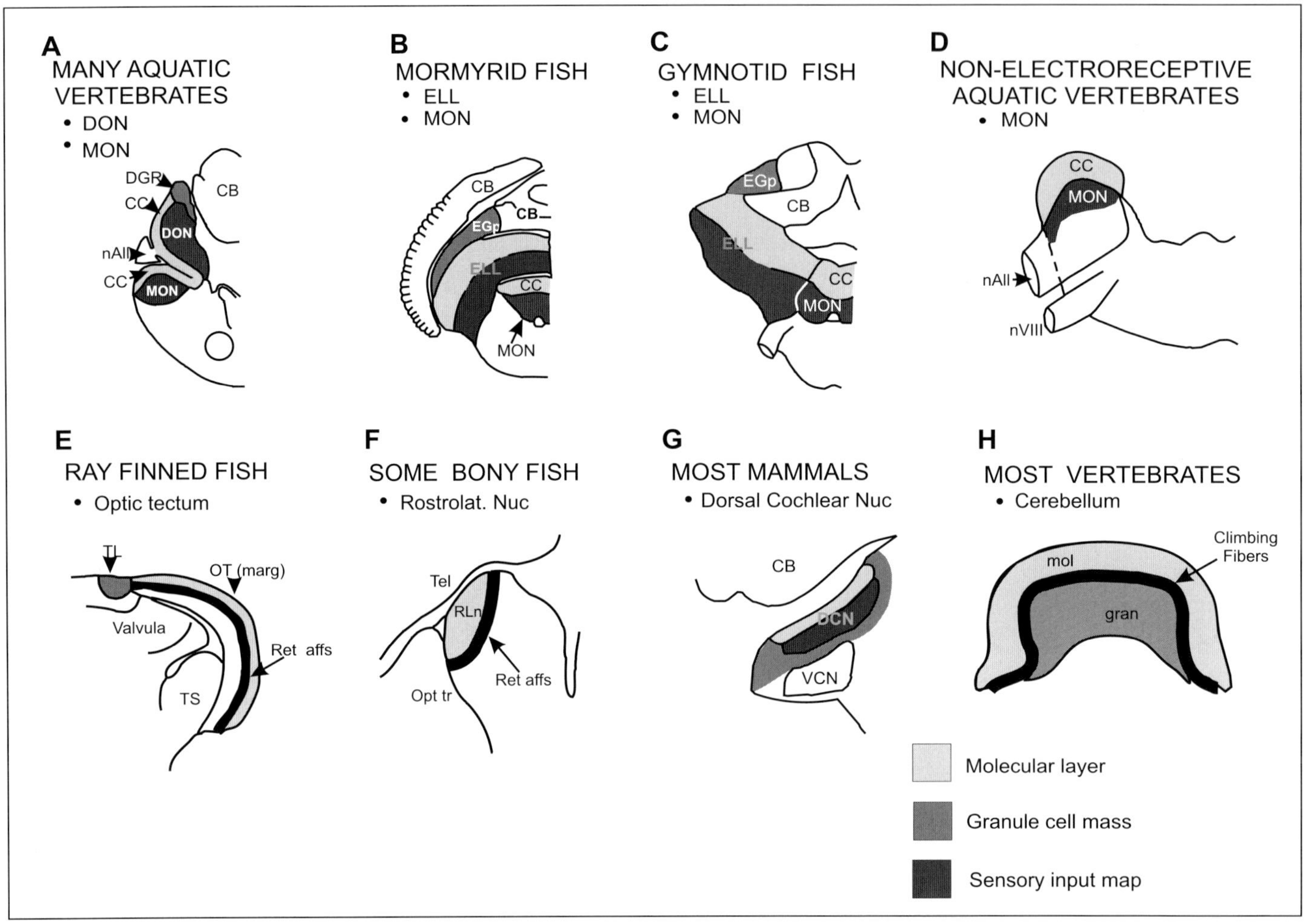

Fig. 2. Cerebellum-like structures in different vertebrate groups. The molecular layer, granule cell mass, and sensory input map are shown in different grayscale, as indicated at the lower right. CB = Cerebellum; CC = cerebellar crest; DCN = dorsal cochlear nucleus; DGR = dorsal granular ridge; DON = dorsal octavolateral nucleus; EGp = eminentia granularis posterior; ELL = electrosensory lobe; gran = granular layer; MON = medial octavolateral nucleus; mol = molecular layer; nAll = anterior lateral line nerve; nVIII = eighth nerve; Opt tr = optic tract; OT (marg) = optic tectum, marginal layer; Ret affs = retinal afferents; RLn = rostrolateral nucleus; Tel = telencephalon; TL = torus longitudinalis; TS = torus semicircularis; VCN = ventral cochlear nucleus.

present in most basal aquatic vertebrates and in some myxinoids; the *dorsal octavolateral nucleus* (DON), present in most of the same basal aquatic vertebrates as the MON, except for the bony fish (neopterygii), where it is entirely absent; the *marginal layer of the optic tectum*, present in all ray finned fish (actinopterygii); the *electrosensory lobe* (ELL), present in a few groups of advanced bony fish (teleosteii); the *rostrolateral nucleus* (RLN) of the thalamus, present in a few groups of bony fish; the *dorsal cochlear nucleus* (DCN), present in almost all mammals; and the *cerebellum* itself, present in all craniates, with the possible exception of the Myxinoidea (fig. 1, 2). Note that the term 'cerebellum-like structure' when used in isolation in this paper includes the cerebellum itself. Similarly, the term 'Purkinje-like cells' includes the Purkinje cells of the cerebellum.

Survey of Cerebellum-Like Structures

General Description of Cerebellum-Like Architecture

A molecular layer is the most universal feature of these structures, and might be taken as a defining feature. The molecular layer contains fine parallel fibers all of which

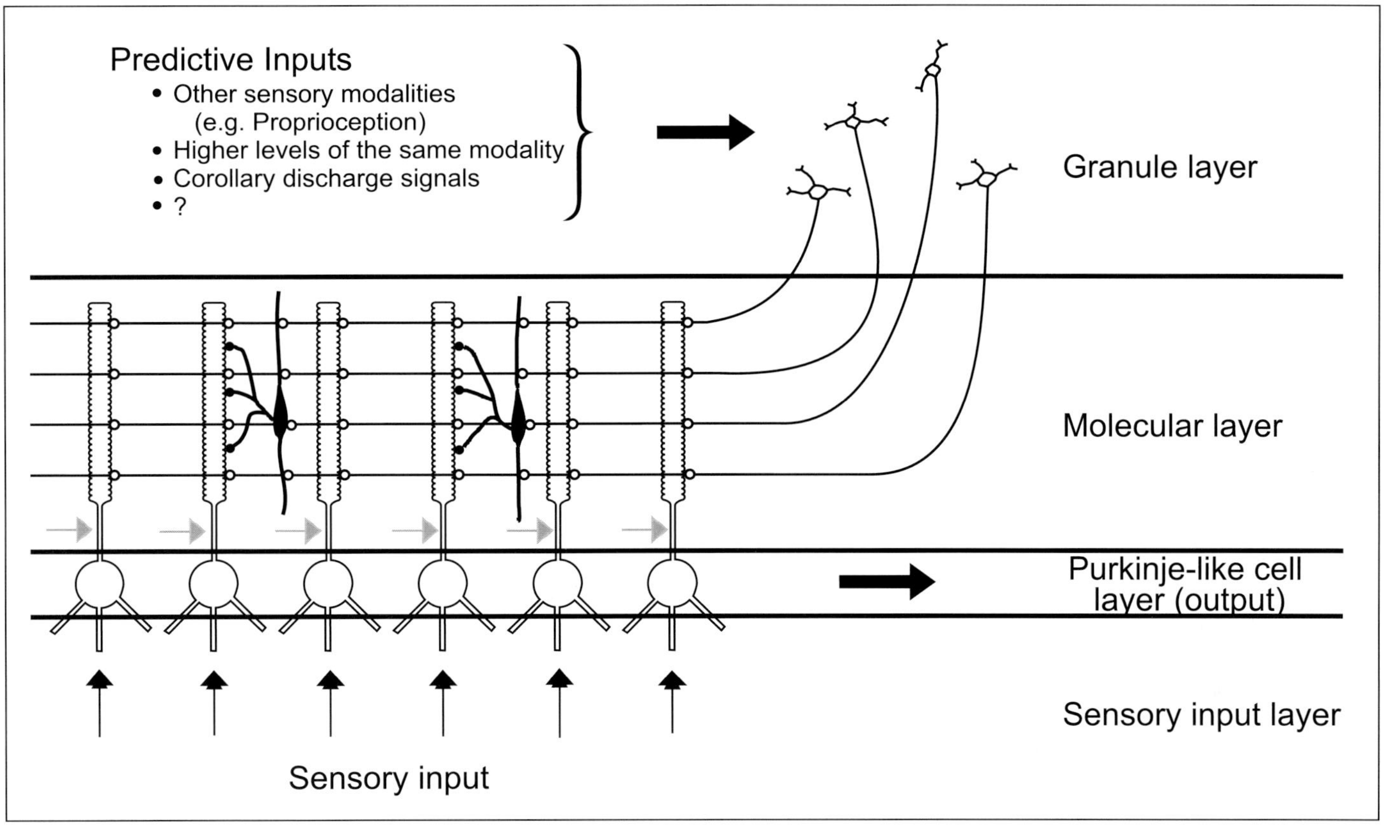

Fig. 3. Schematic drawing showing major features of cerebellum-like structures. Sensory input to the deeper layers may terminate on basal dendrites of Purkinje-like cells (black arrows) or on proximal apical dendrites of these cells (gray arrows).

course in approximately the same direction. Most cerebellum-like structures also have a cell body layer beneath the molecular layer that contains the somas of Purkinje-like cells. These cells have apical dendrites that extend up into an overlying molecular layer and are densely covered with spines. The spines receive synaptic input from parallel fibers (fig. 3). In a few cases, the cell bodies of Purkinje-like cells do not form a distinct layer but are embedded in the molecular layer itself. This occurs in the cerebellum of lampreys and the caudal (vestibulolateral) cerebellar lobe of some fish.

All of the cerebellum-like structures are derived from the alar or sensory plate of the embryo, and most of them can be described as processors of sensory information. Each structure, or in some cases different regions within a structure, processes a particular modality or type of sensory information. Afferent fibers that convey the type of sensory information for which the structure is responsible affect the basal regions of the Purkinje-like cells, either directly or via interneurons in the deeper layers of the structure. The synapses conveying sensory input may be on basilar dendrites (vertical black arrows of fig. 3), on the proximal portion of apical dendrites (horizontal gray arrows of fig. 3), or on both. The afferent fibers usually terminate in a topographically organized manner, providing a map of a peripheral sensory surface such as the retina, the cochlea, or the electroreceptors of the skin.

The parallel fibers in the molecular layer generally arise from a mass of granule cells that is external to the cerebellum-like structure, although in the case of the cerebellum and the dorsal cochlear nucleus granule cells that give rise to parallel fibers are considered a part of the same structure. Neither parallel fibers nor the granule cells that give rise to them have been recorded in vivo in any cerebellum-like structure, and the exact information conveyed by these fibers is not known. Inputs to the granule cells have been described both anatomically and physiologically, however. Thus, the general types of information conveyed by parallel fibers are known in most cases, although the ways in which this information is integrated and encoded by the granule cells are not known. Inputs to the granule cells include signals from sensory modalities other than the one that projects to the deeper layers, signals from higher central nervous system levels of the same sensory modality that projects to the deeper layers, and corollary discharge signals associated with motor commands. All of these parallel fiber signals are ones that are often associated with changes in the sensory modality that the structure is responsible for, allowing the parallel fiber signals to be used as predictors of changes in that modality. The inputs to the granule cells that give rise to parallel fibers are therefore indicated as 'predictive inputs' in figure 3. Other, as yet unidentified, types of predictive input to granule cells may also be present ('?' in fig. 3).

Figure 3 is a simplifying diagram that shows the common features of the different cerebellum-like structures. Each of the actual structures includes a rich variety of additional types of neurons and additional types of input from other central structures. The following descriptions of the different structures also emphasize the common features.

Medial Octavolateral Nucleus

The medial octavolateral nucleus (MON), also known as the intermediate octavolateral nucleus (fig. 2A–D), analyzes sensory input from mechanical lateral line and eighth nerve end organs [McCormick, 1999]. The MON is present in all aquatic craniates that possess mechanical lateral line receptors (fig. 1). Atlantic hagfish (myxinoids) do not have lateral line receptors and do not appear to have a MON [C.B. Braun, personal communication].

Purkinje-like cells of the MON, known as crest cells, have apical dendrites that extend into an overlying molecular layer known as the cerebellar crest. Basilar dendrites of the crest cells are affected by primary afferent fibers from lateral line and eighth nerve end organs [Puzdrowski and Leonard, 1993; McCormick, 1999]. Primary afferents from different sources terminate in different regions of the MON. In the mormyrid fish, for example, fibers from the anterior lateral line nerve, the posterior lateral line nerve, the utricle, the lagena, and the sacculus terminate within separate rostro-caudal strips of the MON [Bell, 1981a].

Parallel fibers of the cerebellar crest descend from an anteriorly located mass of granule cells known as the lateral granule mass in elasmobranchs and the eminentia granularis in other fish. The parallel fibers course in an anterior-to-posterior direction through the cerebellar crest and synapse on the apical dendrites of crest cells. The inputs to these granule cells are not well known but include lateral line primary afferents [Bodznick and Northcutt, 1980], eighth nerve primary afferents [Bell, 1981a; Puzdrowski and Leonard, 1993], input from the spinal cord [Schmidt and Bodznick, 1987], and descending input from higher order lateral line and acoustic centers [Bell, 1981b; Finger and Tong, 1984; Schmidt and Bodznick, 1987; McCormick and Hernandez, 1996; McCormick, 1997].

Dorsal Octavolateral Nucleus

The dorsal octavolateral nucleus (DON; fig. 2A) analyzes sensory input from electroreceptors and is found in aquatic vertebrates including lampreys, many fishes and salamanders [Bullock and Heiligenberg, 1986; Montgomery et al., 1995] (fig. 2). Electroreception is a basal vertebrate sense that may have originated as early as the lateral line or vestibular senses during evolution [Bullock et al., 1983]. The Myxinoidea (hagfish) are not electroreceptive and do not have a DON. Whether electroreception was lost in the evolution of the Myxinoidea, or whether an absence of electroreception is the ancestral craniate condition is not known. Electroreception was clearly lost, however, during the evolution of the bony fish (Neopterygii), and bony fish do not have a DON. Electroreception reappeared during the evolution of a few separate groups of advanced bony fish (Teleosteii), but the more recently derived electroreceptors and associated central structures of teleosts (see below, 'Electrosensory Lobe of Electroreceptive Teleosts') are quite different from those of other aquatic vertebrates.

The DON is located just dorsal to the MON and is similar to the MON in its structure and connections. However, the deeper layers of DON receive primary afferent input from electroreceptors in the skin rather than from mechanical lateral line or eighth nerve receptors [Bodznick and Northcutt, 1980; Puzdrowski and Leonard, 1993]. The afferent projection to DON is topographically organized and forms a map of the electroreceptive surface of the skin. The basilar dendrites of DON crest cells are affected directly or indirectly by primary afferent input, as in the MON, and their apical dendrites extend up into an overlying cerebellar crest, where they are contacted by parallel fibers.

Parallel fibers of the cerebellar crest lying above the DON arise from a mass of granule cells known as the dorsal granular ridge. Anatomical studies show inputs to the dorsal granular ridge from the spinal cord and from other central structures, including ones that appear to be higher-order electrosensory nuclei [Bodznick and Boord, 1986;

Schmidt and Bodznick, 1987]. Inputs to the dorsal granular ridge have also been recorded physiologically in an elasmobranch [Conley and Bodznick, 1994; Hjelmstad et al., 1996]. These inputs fall into three main categories: 1) proprioceptive inputs driven by movements of the body, fins, and respiratory apparatus; 2) electrosensory inputs descending from a higher level electrosensory nucleus; and 3) corollary discharge signals associated with the motor commands that drive the respiratory movements of the fish. Fibers conveying all three types of input discharge in relation to the fish's respiratory cycle, with different fibers being active at different phases of the cycle. Electroreceptors are exquisitely sensitive in elasmobranchs and are strongly affected by the fish's own respiration [Montgomery and Bodznick, 1993]. The descending activity in parallel fibers can therefore be used to predict these cyclic changes in afferent input to the deeper layers of DON (see below, 'Possible Similarity of Function for Cerebellum-Like Structures').

Marginal Layer of the Optic Tectum in Teleosts

Afferent input from the retina terminates in the roof of the mesencephalon in all vertebrates with functional retinas. The region is known as the optic tectum in fish, amphibia, reptiles and birds, and as the superior colliculus in mammals. The optic tectum of actinopterygian (ray-finned) fishes is distinctive in that its outer layers are cerebellum-like [Vanegas et al., 1979; Meek, 1992a].

The most external layer of the actinopterygian tectum is a molecular layer known as the marginal layer (fig. 2E). The marginal layer is small in chondrostean fish (sturgeons and paddle fish), but quite prominent in many holostean and teleost fish. Purkinje-like cells [Type I interneurons of Meek and Schellart, 1978] form a layer deep to the marginal layer. These cells extend their spine-covered apical dendrites up into the marginal layer. Input from the retina maps topographically onto basilar dendrites of the Purkinje-like cells and onto smooth proximal portions of the apical dendrites just above the cell bodies.

The parallel fibers of the marginal layer arise from a medially located granule cell mass known as the torus longitudinalis. The parallel fibers stream laterally away from the torus longitudinalis over the entire surface of the tectum, making synaptic contact with the spines of the Purkinje-like cells.

Granule cells of the torus longitudinalis are affected by signals that could serve to predict the visual input from the retina. Thus, these cells receive a projection from the oculomotor nuclei [Wulliman, 1994] and respond to eye movements in the dark [Northmore et al., 1983], suggesting that they are affected by corollary discharge signals associated with the motor commands that evoke eye movements. Such corollary discharge signals could serve to predict the changes in retinal input evoked by the associated eye movements. The torus longitudinalis cells also receive visual input, possibly from the tectum [Northmore et al., 1983]. One type of visual stimulus will often be associated with another type of visual stimulus. Thus, visual responses of parallel fibers could also serve to predict some aspects of visual input from the retina.

Electrosensory Lobe of Electroreceptive Teleosts

Electroreception is present in four groups of teleosts: Mormyriformes, an order of electric fish from Africa; Gymnotiformes, a superorder of electric fish from South America; Siluriformes, the order of catfish; and Xenomystinae, an African subfamily of the family Notopteridae [Bullock and Heiligenberg,1986]. All of these fish have a cerebellum-like electrosensory lobe (ELL) that receives the afferent input from electroreceptors (fig. 2B, C) [Maler et al., 1981; Braford, 1982; Finger and Tong, 1984; Meek et al., 1999]. Mormyriform and xenomystid fish are closely related taxa of the teleost cohort Osteoglossomorpha (fig. 2), and it is possible that electroreception evolved in a common ancestor but was subsequently lost in non-electroreceptive notopterid fish. Similarly, gymnotiform and siluriform fish are closely related taxa of a different teleost cohort, the Euteleostei, and it is possible that electroreception evolved in a common ancestor of these two euteleost groups; however, electroreception is rare in teleosts, and the many sister groups of the electroreceptive taxa within euteleost or osteoglossid cohorts are not electroreceptive. Thus, electroreception and its central pathways must have evolved independently at least twice within the teleost radiation.

The apical dendrites of ELL Purkinje-like cells extend up into an overlying molecular layer. Their basilar dendrites receive afferent input from electroreceptors, either directly or indirectly via interneurons. The afferent input is topographically organized, forming maps within ELL of the electroreceptive skin surface. The ELLs of mormyriform and gymnotiform fish contain multiple maps. ELLs of the family Mormyridae have three maps, each receiving a different type of primary afferent [Bell and Szabo, 1986]. Gymnotiform ELLs have four maps, but the maps are formed in a different manner. Three of the maps in the gymnotiform ELL are formed by branching of the same afferent fibers. A fourth map receives a different type of electroreceptor afferent [Carr and Maler, 1986].

The basic cerebellum-like structure of ELL is similar in mormyrid and gymnotiform fish, but there are also important differences, in addition to the difference in how maps are formed. Purkinje-like cells in the mormyrid ELL are inhibitory interneurons that end locally on efferent neurons, which in turn convey the electrosensory information to the mesencephalon. In this regard, mormyrid Purkinje-like cells are like true Purkinje cells of the teleost cerebellum. Teleost Purkinje cells are also inhibitory interneurons that end locally on efferent neurons, which are equivalent to cells of the cerebellar nuclei in mammals [Meek, 1992b]. In contrast, Purkinje-like cells of the gymnotid ELL are not inhibitory and are themselves efferent cells projecting to the mesencephalon [Carr and Maler, 1986]. These and other differences between the ELLs of the two groups are consistent with their independent evolutionary origins.

The parallel fibers of the ELL molecular layer arise from the eminentia granularis posterior (EGp), a mass of granule cells that is continuous with the anterior eminentia granularis from which parallel fibers of the cerebellar crest arise in the same fish. The inputs to EGp have been most extensively studied in mormyrid and gymnotid fish. In mormyrid fish these inputs include corollary discharge signals associated with the motor command that elicits the electric organ discharge, proprioceptive signals associated with bending of the body or the fins, and descending input from higher levels of the electrosensory system [Bell et al., 1992]. Anatomical studies show that the mormyrid EGp also receives extensive input from the spinal cord via a tract that appears to be equivalent to the ventral spinocerebellar tract in mammals [Szabo et al., 1979]. The ventral spinocerebellar tract in mammals conveys corollary discharge information associated with motor activity generated in the spinal cord [Arshavskii et al., 1972]. Thus it is possible that the mormyrid EGp might also receive corollary discharge information associated with ordinary motor activity. The EGp of gymnotid fish also receives proprioceptive input and descending input from higher levels of the electrosensory system [Bastian, 1986]. All of these different types of signals are relayed to ELL as parallel fiber inputs, where they can serve to predict changes in the afferent activity from electroreceptors arriving in the deeper layers (see below, 'Possible Similarity of Function for Cerebellum-Like Structures').

Rostrolateral Nucleus of the Thalamus in Some Holostean and Teleost Fish

The rostrolateral nucleus (RLN, fig. 2F) of the thalamus is a small cerebellum-like structure found in the thalamus of a few widely scattered neopterygian fish (fig. 1). The nucleus has been identified in 6 of the 16 different taxa in which it has been investigated [Butler and Saidel, 1992; Saidel and Butler, 1997]. The distribution in different fish is sporadic, and the structure is not present in sister groups of taxa in which it is present. The Purkinje-like cells of RLN do not have basilar dendrites but receive a direct input from the retina on the most proximal part of their apical dendrites (as shown by gray arrows in fig. 3). This input appears to be retinotopically organized [W.M. Saidel, personal communication]. The more distal apical dendrites are covered with spines and receive parallel fiber input from the torus longitudinalis. It is not known whether parallel fibers in the marginal layer of the tectum and the RLN arise from the same granule cells or from separate granule cells in the torus longitudinalis. The RLN has not yet been studied physiologically.

Dorsal Cochlear Nucleus of Mammals

All mammals possess a dorsal cochlear nucleus (DCN; fig. 2G). The DCN is laminated and cerebellum-like in marsupials and eutherian mammals but not in monotremes [Cant, 1992; Johnson et al., 1994; Nieuwenhuys et al., 1997]. The size of DCN varies greatly in different taxa, being rather small in humans and very large in some rodents [Merzenich et al., 1973]. Functional correlates of this variation in size are not known. Primary afferent fibers from the cochlea terminate in a tonotopically organized manner in the deeper cell layers of the structure (fig. 1G). An overlying molecular layer contains parallel fibers that arise from granule cells located around the margins of the nucleus. The parallel fibers course at right angles to the isofrequency bands in the deeper layers. Thus, each parallel fiber passes through different frequency-specific regions of DCN.

The cells that are most similar to Purkinje cells in the DCN are the 'cartwheel' cells [Cant, 1992]. The dendrites of these cells are restricted to the molecular layer. Cartwheel cells are inhibitory interneurons that terminate locally on efferent neurons [Oertel and Wu, 1989], being similar in this respect to Purkinje cells of the teleost cerebellum and Purkinje-like cells of the mormyrid ELL. Cartwheel cells are different from most of the previously described Purkinje-like cells in that they do not have basilar dendrites and may not be directly affected by primary afferent input.

There is another cell type in the DCN, however; the fusiform cell is more similar to the previously described Purkinje-like cells in that it has both apical and basilar dendrites [Oertel and Wu, 1989; Cant, 1992]. The cell bodies of fusiform cells are in the deeper cell layer. Their spine-covered apical dendrites extend up into the molecular layer where they are contacted by parallel fibers. Their basilar dendrites are in the cell layer where they are contacted by primary afferent fibers from the cochlea. The fusiform cells are efferent cells and project to the mesencephalon.

Various types of input to the granule cells of DCN have been identified including somatosensory input [Weinberg and Rustioni, 1987] and descending input from higher levels of the brain. The descending input comes from the auditory cortex [Weedman and Ryugo, 1996], somatosensory cortex [Wolff and Kunzle, 1997], and the inferior colliculus [Caicedo and Herbert, 1993]. Somatosensory input comes from the dorsal column nuclei as well as from the somatosensory cortex. Somatosensory input from the pinna has a particularly strong effect on cells of DCN [Kanold and Young, 2001], with proprioceptive input associated with pinna movement being much more effective than cutaneous input. The DCN granule cells also receive somatosensory input from the neck and forelimbs. Movements of the animal's pinna, head, or body will have strong and predictable effects on how the cochlea responds to an external sound source. Similarly, different sounds often occur together or in sequence so that one sound can predict another sound, as in a remembered melody. Thus, the somatosensory and auditory signals conveyed by the parallel fibers of DCN can serve to predict changes in afferent activity from the cochlea that arrives at the deeper layers.

Cerebellum

All vertebrates appear to have a cerebellum (fig. 2H), but the existence of a cerebellum is controversial in hagfish, with some anatomists saying it is present [Larsell, 1967] and others saying that it is not [Nieuwenhuys et al., 1997]. Although most anatomists conclude that a cerebellum is present in lampreys, this cerebellum is small and difficult to distinguish [M.A. Pombal, personal communication], as it is made up of a few scattered cells in the rhombic lip that are embedded among commissural fibers and what may be parallel fibers. However, the MON with its overlying cerebellar crest is clear in Pacific hagfish (Eptatretidae) and in lampreys. The DON with its cerebellar crest is also clear in lampreys. Thus, these cerebellum-like structures, the MON and DON, probably evolved as early as, or even earlier than, the cerebellum itself.

The traditional view that the cerebellum is an organ of motor control may be only partially correct. Paulin [1993] has suggested that the cerebellum might be better viewed as a sensory processor that analyzes and processes sensory information for subsequent use by the motor system. Several lines of evidence suggest a sensory role. First, the regions of the cerebellum that are relatively enlarged in different animals are those that receive input from the most highly developed senses in those animals. The enlarged regions of the cerebellum in echolocating bats and cetaceans, for example, are those that receive auditory input. Second, the lack of a sharp border between the granule cell masses of some cerebellum-like structures that are clearly sensory and the granule cell layer of the cerebellum is also consistent with a sensory role for the cerebellum. Thus, the eminentia granularis of teleosts, the lateral granule mass of elasmobranchs, and the granule cells of DCN all merge with the granule layers of the different cerebellums. Finally, embryology suggests a sensory role for the cerebellum as the cerebellum is derived from the sensory (alar) plate of the neural tube. Emphasis on the role of the cerebellum as a sensory processor, analyzing such input for subsequent use by the motor system, makes it possible to group the cerebellum with the other cerebellum-like structures, all of which are clearly sensory.

The most obvious difference between the histology of the cerebellum and that of the other cerebellum-like structures is the climbing fiber input to cerebellar Purkinje cells. Each Purkinje cell receives input from a single climbing fiber that makes a very large number of synaptic contacts on the same cell. Such climbing fibers have been observed morphologically in most vertebrate groups, including elasmobranchs [Alvarez-Otero et al., 1993], teleosts [Meek and Nieuwenhuys, 1991], birds [Cajal, 1952], amphibians [Sotelo, 1969], reptiles [Hillman, 1969], and mammals [Cajal, 1952]. Physiologically, a single action potential in the climbing fiber evokes a large, all-or-none, excitatory postsynaptic potential (EPSP) in the Purkinje cell that in turn evokes postsynaptic spikes. Such climbing fiber responses have been observed in all vertebrates in which Purkinje cells have been recorded, including teleosts [Kotchabhakdi, 1976; V.Z. Han, personal communication], amphibians [Naciemiento, 1969], reptiles [Llinas and Nicholson, 1969], and mammals [Eccles et al., 1967]. Such strong synaptic input from a single presynaptic fiber has not been observed either morphologically or physiologically in any cerebellum-like structure other than the cerebellum.

Climbing fibers terminate on the smooth branches of Purkinje cells, not on the spiny branchlets where parallel fibers terminate. The smooth branches extend throughout the molecular layer in mammals, but in all other vertebrate classes the smooth dendrites are confined to a deeper portion of the molecular layer. Thus, the climbing fibers terminate on the smooth, proximal portions of Purkinje cell dendrites in non-mammals (as illustrated by gray arrows in fig. 3).

Climbing fiber input is considered here as analogous, at least, to the sensory input to deep layers in the other cerebellum-like structures. Climbing fibers arise from the inferior olive at the base of the brain. But cells of the inferior olive are derived from the alar plate (like the cerebellum) and migrate to their ventral location during development. Different parts of the inferior olive receive input from different sensory systems, and climbing fibers to different parts of the cerebellum respond with great sensitivity and specificity to sensory stimuli, including visual [Maekawa and Simpson, 1972], auditory [Mortimer, 1975], vestibular [Precht et al., 1976; Barmack and Shojaku, 1992], and somatosensory [Robertson, 1985] stimuli.

The inferior olive is certainly much more than a sensory relay. This additional complexity is indicated by the large projections to the inferior olive from the cerebral cortex, the large projections from inhibitory neurons of the cerebellar nuclei [Fredette and Mugnaini, 1991], the intrinsic rhythmicity of the inferior olive [Llinas and Yarom, 1986], and the gating of climbing fiber responses to sensory input by motor command signals [Gellman et al., 1985; Lidierth and Apps, 1990]. Nevertheless, it is clear that climbing fibers in many parts of the cerebellum convey precise sensory information to the smooth dendrites of Purkinje cells.

Granule cells of the cerebellum receive input from a rich variety of sources, including motor command related signals, sensory signals from many different modalities, and signals from different regions of the cerebral cortex relayed through the pons. These signals are conveyed by parallel fibers to Purkinje cells where they can be used to predict the input conveyed by climbing fibers to the proximal dendrites (see below, 'Possible Similarity of Function for Cerebellum-Like Structures').

Additional Anatomical Similarities

The previous section emphasized the anatomical features that are shared by all of the different cerebellum-like structures. These included a molecular layer with parallel fibers that course in the same direction and Purkinje-like cells that receive the parallel fiber input on the spines of their apical dendrites and sensory input on their basilar dendrites or on the proximal portion of their apical dendrites. Additional similarities are also present which may be shared by only a subset of the different cerebellum-like structures or have not yet been looked for in all of the structures. These include similar cell types, proteins, and connections.

Cell types within the granule cell masses that give rise to the parallel fibers appear to be very similar in several of the cerebellum-like structures. Granule cells in the cerebellum [Cajal, 1952], in the DCN [Kane, 1974], in the eminentia granularis [where the parallel fibers of ELL and MON arise; Bell, unpublished observations in the mormyrid] all have the classic granule cell shape of a small round cell bodies with 3 to 6 thin dendrites extending out radially and terminating in claw-like postsynaptic structures. Large, GABAergic Golgi cells are similarly distributed throughout the granule layer in all these structures. An unusual and only recently identified cell, the unipolar brush cell, is also present in the cerebellum [Floris et al., 1994], DCN [Floris et al., 1994], and eminentia granularis [Bell, unpublished observations], but is not present in structures that are not cerebellum-like. Finally, GABAergic stellate cells are present in the molecular layer of most cerebellum-like structures [Montgomery et al., 1995] (fig. 3).

Immunocytochemistry and in situ hybridization show that the same or similar genes are expressed in different cerebellum-like structures. This has been especially studied with regard to the cerebellum and the DCN, where antibodies to many proteins stain both Purkinje cells and DCN cartwheel cells [Berrebi et al., 1990]. Unipolar brush cells in both structures [Floris et al., 1994] and in the eminentia granularis of mormyrids [Bell, unpublished observations] stain with antibodies to calbindin, calretinin, and metabotropic glutamate receptors 1 and 2/3. Immunocytochemistry shows that the dendrites of Purkinje cells [Maeda et al., 1990], DCN cartwheel cells [Ryugo et al., 1995] and Purkinje-like cells of the teleost optic tectum marginal layer [Berman et al., 1995; Bell, unpublished observations] express high levels of 1,4,5-phosphate (IP3) receptors, but that the dendrites of Purkinje-like cells of the ELL do not [Berman et al., 1995]. The NR1 subunit of the NMDA receptor is present in Purkinje-like cells of the teleost ELL [Berman et al., 2001], the teleost MON [Bell, unpublished observations], the teleost optic tectum [Bell, unpublished observations], and the DCN [Petralia et al., 1996], but cerebellar Purkinje cells do not show such staining.

The external connections of some of the different types of cerebellum-like structures can be quite similar. Several of the cerebellum-like structures that receive sensory input from octavolateral receptors (eighth nerve, lateral line, or electroreceptor end organs) project bilaterally to mesencephalic nuclei via the lateral lemniscus. Such structures include the DON, MON, ELL, and DCN. In contrast, the cerebellum, parts of which receive primary afferent input from the eighth nerve, has a quite different pattern of projection. The similar afferent connections, from the retina and the torus longitudinalis, of the RLN and the optic tectum marginal layer have already been mentioned.

Possible Similarity of Function for Cerebellum-Like Structures

The ELL of mormyrid fish, the ELL of gymnotid fish, and the DON of elasmobranch fish have all been shown to act as adaptive sensory processors [Bell, 1982; Bodznick, 1993; Bastian, 1995]. Each of these structures can generate a negative image of predicted electrosensory input patterns after a period of association between predictive signals conveyed by parallel fibers and a consistent pattern of electrosensory input to the deeper layers. Addition of the negative image of predicted input to the actual input removes or minimizes predicted features of the sensory inflow. Such removal allows novel sensory features to stand out and makes better use of the information processing and storage capacity of the brain. Adaptive sensory processing in these structures has been reviewed elsewhere [Bell et al., 1997a; Bell, 2001] and is described only briefly here.

The electric organ discharge (EOD) of mormyrid electric fish affects the ELL in two distinct ways – via primary afferent fibers from electroreceptors excited by the EOD and via an electric organ corollary discharge (EOCD) signal associated with the motor command that evokes the EOD. The two signals can be separated by giving curare to the fish. Curare blocks the EOD itself but the EOD motor command continues to be emitted spontaneously ('fictive discharging'). The effect of the electric organ corollary discharge (EOCD) on Purkinje-like cells of ELL is transformed by a few minutes of pairing with an electrosensory stimulus to the receptive field of the cell (fig. 4A). Such pairing results in an EOCD response that was not present before the pairing and which is a negative image of the previously paired sensory response. Similar experiments show that the negative image is specific to the polarity of the paired sensory response (excitation vs. inhibition) and to its timing relative to the EOD motor command. The EOCD-evoked negative image is also specific to the spatial pattern of the paired sensory response as the sensory stimulus must affect the cell, that is, be within the cell's receptive field, in order to be effective [Bell, 1982].

Similar negative image formation occurs in the ELL of gymnotid electric fish after pairing electrosensory stimuli with tail bending [Bastian, 1995] (fig. 4B), and in the DON of elasmobranch fish after pairing electrosensory stimuli with respiratory ventilation [Bodznick, 1993] (fig. 4C). Negative images in the gymnotid ELL and in the elasmobranch DON are specific to the polarity, the timing, and the spatial distribution of the sensory response, as in the mormyrid ELL. The canceling of a predicted sensory response by the addition of a negative image of that response is shown particularly clearly in figure 4C, where coupling to the ventilatory rhythm reduces a large initial response ('0 min') to almost nothing ('25 min').

A variety of different predictive signals can be used to evoke negative images in the gymnotid ELL and the elasmobranch DON. Thus, both proprioceptive signals from the trunk and whole body electrosensory signals can evoke negative images of electrosensory responses with which they have been paired in the gymnotid ELL [Bastian, 1999]. Similarly, proprioceptive signals, corollary discharge signals, and whole body electrosensory signals are all effective as predictive signals in the elasmobranch DON [Bodznick et al., 1999].

Plasticity has been demonstrated at synapses between parallel fibers and Purkinje-like cells in the ELL of mormyrids [Bell et al., 1997b], the ELL of gymnotids [Bastian, 1999], and the DON of elasmobranchs [Bodznick et al., 1999]. Both depression and potentiation of synaptic strength are present. Several lines of evidence [Bell et al., 1997a] as well as modeling studies [Nelson and Paulin, 1995; Roberts and Bell, 2000] strongly suggest that plasticity at this synapse is responsible for the negative images of predicted sensory responses.

All of the cerebellum-like structures are characterized by parallel fibers conveying signals that can predict sensory input to the deeper layers and by Purkinje-like cells that receive both parallel fiber and sensory inputs. Thus, the results obtained in mormyrids, gymnotids and elasmobranchs suggest the hypothesis that similar adaptive processing might occur in other cerebellum-like structures. In the dorsal cochlear nucleus, for example, parallel fibers conveying information about movements of the pinna or of the head could elicit negative images of the expected changes in auditory input, a possibility that is supported by the recent discovery of associative synaptic plasticity at parallel fiber synapses in the dorsal cochlear nucleus [Fujino and Oertel, 2002]. The creation of negative images of predicted inputs and the removal of predictable

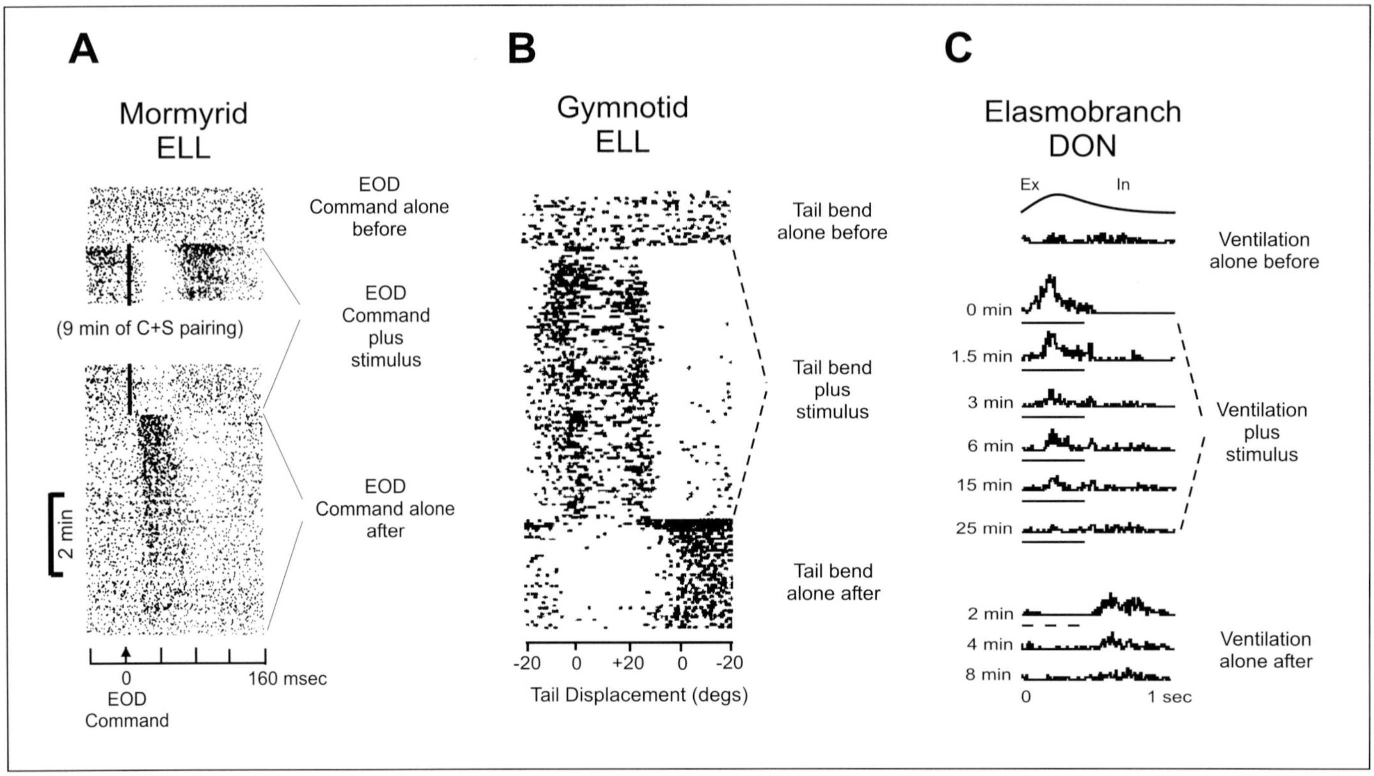

Fig. 4. Formation of negative images of predicted sensory responses in three different cerebellum-like structures. **A** Raster display of the responses of a cell in the mormyrid ELL. Each dot represents an action potential. The EOD motor command occurs at time 0. The command initially has no effect on the cell. An electrosensory stimulus (vertical black line) that evokes a pause-burst response is then paired with the command. After several minutes of pairing, the stimulus is turned off and a response to command alone is revealed which was not present before the pairing and which is a negative image of the previously paired sensory response [from Bell, 1986]. **B** Raster display of responses of cell in the gymnotid ELL. The tail is moved back and forth passively. Each row of dots shows response to one cycle of movement. Initially the tail bend has no effect on the cell. An electrosensory stimulus that evokes a burst-pause is then delivered in phase with the movement. The electrosensory stimulus is turned off after several minutes of pairing, revealing a response to tail bending alone that was not present before the pairing and which is opposite to the previously paired sensory response [from Bastian, 1995]. **C** Histogram display of responses of a cell in the elasmobranch DON. Initially the cell does not respond to the exhalation ('Ex') – inhalation ('In') ventilatory cycle of the fish (top histogram). An electrosensory stimulus that evokes a burst-pause is then delivered in phase with the ventilatory cycle. The response to ventilation plus the electrosensory stimulus decreases during 25 min of pairing. Turning off the electrosensory stimulus after pairing reveals the presence of a response to ventilation alone which was not present before and which is opposite to the previously paired sensory response [from Bodznick, 1993].

features would be useful in many systems, and selection for the ability to carry out such processing could have contributed to the evolution of cerebellum-like structures in different systems.

The cerebellum itself might carry out a similar type of adaptive processing [Bell et al., 1997a; Devor, 2000]. Plastic change has been demonstrated at the parallel fiber to Purkinje cell synapse following the pairing of parallel fiber and climbing fiber inputs [Ito, 1984, 2001]. Moreover, parallel fibers in many regions of the cerebellum convey signals that could be used to predict the climbing fiber inputs to the same cerebellar region. Climbing fibers are commonly viewed as signaling an undesirable condition, such as an error in motor performance. If synaptic plasticity results in parallel fiber activity eliciting a negative image of expected climbing fiber input, and if the Purkinje cells are connected to the rest of the brain so that a pause in their activity opposes whatever it was that evoked the climbing fiber input, then the cerebellum would serve to minimize the undesirable conditions signaled by climbing fibers. The described process is only a restatement of Ito's theory of motor learning [Ito, 1984], but it places the cerebellum within the framework of the other cerebellum-like structures.

Evolutionary and Developmental Origins

The concept of homology is of central importance for understanding the evolutionary origins of similarities among different structures in different taxa. The concept has been used in various ways, but will be used here in the sense of historical or phylogenetic homology [Bolker and Raff, 1996; Futuyma, 1998; Butler and Saidel, 2000]. In this usage, a character is considered homologous across different taxa if the taxa have inherited the character from a common ancestor that also had the character. Such homology requires a congruence between the distribution of the character in the different taxa and their phylogeny, as determined by analysis of all the other characters and the fossil record. If the presumed common ancestor of two taxa with a similar character did not have the character, then the character is not homologous. There must also be a continuity of information. If the character is present in two different taxa and also in their common ancestor, but disappeared and then reappeared during the evolution of one of the taxa, then the character is not homologous.

The cerebellums of different vertebrates are clearly homologous according to this phylogenetic definition. All vertebrates from lampreys to mammals have a cerebellum, and parsimony therefore implies that the cerebellum was present in the common ancestor. Thus, the similarity among vertebrate cerebellums is best explained by homology. In addition, similarity of the same cerebellum-like structure across different taxa can, in most cases, also be best explained by homology. This is true for the MONs of aquatic vertebrates and myxinoids, the DONs of many electroreceptive aquatic vertebrates, the marginal layer of the optic tectum of ray finned fish, and the DCN of marsupial and eutherian mammals. In each case, the phylogeny indicates that the group that possesses the cerebellum-like structure had a common ancestor, and that similarity of the structure across different taxa can be explained by evolution from that common ancestor.

The similarities among ELLs of different electroreceptive teleosts or among the RLNs of different bony fish cannot be explained by homology, however. The osteoglossid and euteleost taxa with ELLs are widely separated from each other in phylogeny and the many sister groups of taxa with ELLs do not have ELLs. Parsimony therefore indicates that the common ancestor of osteoglossid and euteleost cohorts did not have an ELL either. Moreover, the ancestor of teleosts is presumed to have been a holostean fish and holosteans do not have an ELL. The different taxa with RLNs are also widely separated from each other in phylogeny and most of their sister groups do not have the nucleus. Parsimony therefore indicates that the common ancestor of taxa with an RLN did not have the structure.

Similarities among the different types of cerebellum-like structures also cannot be explained by homology. There is no evidence for an ancestral cerebellum-like structure from which the cerebellum, MON, DON, marginal layer of the tectum, ELL, RLN, and DCN all evolved. Indeed, there is no evidence for an ancestral structure from which even two of the presently existing cerebellum-like structures evolved.

How then can we explain the similarities among the different cerebellum-like structures? Structural similarities that are not due to homology are referred to as examples of 'homoplasy'. Homoplasy is considered to have three possible origins: *parallel evolution*, in which two structures evolved independently but in a similar manner under similar selective pressure from the same ancestral structure; *evolutionary reversal*, in which a structure that is present in the common ancestor of two taxa disappeared and then reappeared in one or both taxa; and *convergence*, in which two structures evolve toward similarity via independent evolutionary paths because of similar selective pressure [Futuyma, 1998].

None of these explanations are really satisfactory as explanations of the similarities among cerebellum-like structures. *Parallel evolution* could have played a role in a few cases perhaps. Thus, it is possible that the DCN evolved from part of the cerebellum of an ancestral mammal by invasion of auditory nerve fibers, or that the ELLs of electroreceptive teleosts evolved from the MONs of ancestral non-electroreceptive teleosts by invasion of electroreceptor afferents into part of the MON. But there is no evidence for these scenarios, such as the existence of intermediate forms in which the cerebellum and the DCN, or the MON and the ELL, are not structurally distinct. More generally, the lack of plausible common ancestors for the different types of cerebellum-like structures, as described previously, also argues against parallel evolution as an explanation of the similarities among these structures. *Evolutionary reversal* might be invoked to explain the ELLs of teleosts as a reappearance of the electroreceptive DON that is present in more basal vertebrates. However, such a hypothesis is rather empty in the absence of more information or a mechanism, and could not, in any case, explain the similarities among other cerebellum-like structures.

Convergence resulting from similar selective pressures could have played a role in the evolution of cerebellum-like structures. This possibility was suggested in the preceding section where cerebellum-like structures were shown to be capable of removing predictable features from the sensory inflow, such a role being of great value for any sensory system in any organism. Although cerebellum-like structures might be well suited to this task, they are not uniquely suited. Other, quite different, neural structures, such as the cerebral cortex, can probably carry out similar types of processing [Singer, 1995]. In addition, it is difficult to imagine how all the striking similarities in these structures could have arisen through completely independent evolutionary processes. Something more than convergence seems to be needed to explain the similarities.

That something more is probably a shared genetic-developmental program present in all craniates that once activated can lead to the generation of a cerebellum-like structure. The many similarities in circuitry, cell morphology, immunocytochemistry, and connectivity suggest the presence of such a shared genetic program, and findings from experimental embryology support the possibility. For example, ectopic cerebellum-like structures will develop in the forebrain or midbrain of a chick embryo if beads are coated with fibroblast growth factor 8 and placed at those sites in the embryo [Martinez et al., 1999]. Similarly, cerebellar tissue will develop ectopically in the midbrain and forebrain of a mouse embryo with a genome that is Otx1+/– and Otx2+/– (*Drosophila* orthodenticle protein, a transcription factor) [Acampora et al., 1997]. Finally, different independent mutations in mice that cause Purkinje cell degeneration also cause degeneration of cartwheel cells in the DCN [Berrebi et al., 1990].

All of the cerebellum-like structures are derived embryologically from the somatosensory portion of the alar plate, and it is possible that the genetic program leading to the development of cerebellum-like structures can only be activated within that portion of the embryo. It is likely too, that cerebellum-like structures that are very similar, such as the DON and the MON or the cerebellum and the DCN, share more of their genetic-developmental machinery than do structures which are rather different, such as the cerebellum and the RLN.

Developmental expression of a similar set of genes in different brain regions, or in different taxa with no common ancestor that also expresses those genes, might be described as convergence in developmental gene expression [Wray, 2002]. Such convergence in gene expression would appear to explain the convergence in histological organization shown by the different cerebellum-like structures. More generally, 'syngeny' is a useful term that has been introduced to describe similarities of structure which are due to a shared genetic-developmental program, whether phylogenetic homology is present or not [Butler and Saidel, 2000]. Syngeny can be used to describe structures in the same organism, such as the forelimb and hindlimb, that are similar and that share some of the same genetic machinery during development but which are not homologous. The ELLs of electroreceptive teleosts, the RLNs of various bony fish, and the different types of cerebellum-like structures are other examples of structures that appear to be syngenous but not homologous. The cerebellums of different vertebrates, however, are examples of structures which are both syngenous and homologous.

In conclusion, the similarities among the different types of cerebellum-like structures can be best explained by convergence rather than by homology or parallel evolution – convergence as constrained by the available genetic-developmental programs and by the requirements for a certain type of information processing.

Acknowledgments

A sincere thank you to all the people who provided information and comments during the writing of this paper: David Bodznick, Christopher Braun, Catherine Carr, Bernd Fritzsch, Richard Hawkes, Catherine McCormick, Johannes Meek, Enrico Mugnaini and William Saidel. Thanks too to Heather Eisthen, Kiisa Nishikawa, and Karger Publishers for a fine symposium on convergence and for inviting me to participate. This work was supported by grants from the National Institute of Mental Health (MH49792 and MH60996).

References

Acampora, D., V. Avantaggiato, F. Tuorto, and A. Simeone (1997) Genetic control of brain morphogenesis through Otx gene dosage requirement. Development, *124:* 3639–3650.

Alvarez-Otero, R., S.D. Regueira, and R. Anadon (1993) New structural aspects of the synaptic contacts on Purkinje cells in an elasmobranch cerebellum. J. Anat., *182:* 13–21.

Arshavskii, Y.I., M.B. Berkinblit, I.M. Gel'fand, G.N. Orlovskii, and O.I. Fukson (1972) Activity of the neurones of the ventral spinocerebellar tract during locomotion of cats with deafferentated hind limbs. Biofizika, *17:* 1112–1118.

Barmack, N.H., and H. Shojaku (1992) Vestibularly induced slow oscillations in climbing fiber responses of Purkinje cells in the cerebellar nodulus of the rabbit. Neuroscience, *50:* 1–5.

Bastian, J. (1986) Electrolocation: Behavior, anatomy, and physiology. *In* Electroreception (ed. by T.H. Bullock and W. Heiligenberg), Wiley, New York, pp. 577–612.

Bastian, J. (1995) Pyramidal-cell plasticity in weakly electric fish: A mechanism for attenuating responses to reafferent electrosensory inputs. J. Comp. Physiol., *176:* 63–78.

Bastian, J. (1999) Plasticity of feedback inputs in the apteronotid electrosensory system. J. Exp. Biol., *202:* 1327–1337.

Bell, C.C. (1981a) Central distribution of octavolateral afferents and efferents in a teleost (Mormyridae). J. Comp. Neurol., *195:* 391–414.

Bell, C.C. (1981b) Some central connections of medullary octavolateral centers in a mormyrid fish. *In* Hearing and Sound Communication in Fishes (ed. by R.R. Fay, A.N. Popper and W.N. Tavolga), Springer, New York, pp. 383–392.

Bell, C.C. (1982) Properties of a modifiable efference copy in electric fish. J. Neurophysiol., *47:* 1043–1056.

Bell, C.C. (1986) Duration of plastic change in a modifiable efference copy. Brain Res., *369:* 29–36.

Bell, C.C. (2001) Memory-based expectations in electrosensory systems. Curr. Opin. Neurobiol., *11:* 481–487.

Bell, C.C., and T. Szabo (1986) Electroreception in mormyrid fish: Central anatomy. *In* Electroreception (ed. by T.H. Bullock and W. Heiligenberg), Wiley, New York, pp. 375–421.

Bell, C., D. Bodznick, J. Montgomery, and J. Bastian (1997a) The generation and subtraction of sensory expectations within cerebellum-like structures. Brain Behav. Evol., *50:* 17–31.

Bell, C.C., K. Grant, and J. Serrier (1992) Corollary discharge effects and sensory processing in the mormyrid electrosensory lobe: I. Field potentials and cellular activity in associated structures. J. Neurophysiol., *68:* 843–858.

Bell, C.C., V.Z. Han, S. Sugawara, and K. Grant (1997b) Synaptic plasticity in a cerebellum-like structure depends on temporal order. Nature, *387:* 278–281.

Berman, N., R.J. Dunn, and L. Maler (2001) Function of NMDA receptors in a feedback pathway of the electrosensory system. J. Neurophysiol., *86:* 1612–1621.

Berman, N.J., M.T. Hincke, and L. Maler (1995) Inositol 1,4,5-trisphosphate receptor localization in the brain of a weakly electric fish *(Apteronotus leptorhynchus)* – with emphasis on the electrosensory system. J. Comp. Neurol., *361:* 512–524.

Berrebi, A.S., J.I. Morgan, and E. Mugnaini (1990) The Purkinje cell class may extend beyond the cerebellum. J. Neurocytol., *19:* 643–654.

Bodznick, D. (1993) The specificity of an adaptive filter that suppresses unwanted reafference in electrosensory neurons of the skate medulla. Biol. Bull., *185:* 312–314.

Bodznick, D., and R.L. Boord (1986) Electroreception in chondrichthyes: Central anatomy and physiology. *In* Electroreception (ed. by T.H. Bullock and W. Heiligenberg), Wiley, New York, pp. 225–256.

Bodznick, D. and R.G. Northcutt (1980) Segregation of electro- and mechanoreceptive inputs to the elasmobranch medulla. Brain Res., *195:* 313–321.

Bodznick, D., J.C. Montgomery, and M. Carey (1999) Adaptive mechanisms in the elasmobranch hindbrain. J. Exp. Biol., *202:* 1357–1364.

Bolker, J.A., and R.A. Raff (1996) Developmental genetics and traditional homology. BioEssays, *18:* 489–494.

Braford, M.R. (1982) African, but not Asian, notopterid fishes are electroreceptive: Evidence from brain characters. Neuroscience Lett., *32:* 35–39.

Bullock, T.H., and W. Heiligenberg (eds.) (1986) Electroreception. Wiley, New York.

Bullock, T.H., D.A. Bodznick, and R.G. Northcutt (1983) The phylogenetic distribution of electroreception: Evidence for convergent evolution of a primitive vertebrate sense modality. Brain Res. Rev., *6:* 25–46.

Butler, A.B., and W.M. Saidel (1992) Tectal projection to an unusual nucleus in the diencephalon of a teleost fish, *Pantodon buchholzi.* Neurosci. Lett., *145:* 193–196.

Butler, A.B., and W.M. Saidel (2000) Defining sameness: Historical, biological, and generative homology. BioEssays, *22:* 846–853.

Caicedo, A., and H. Herbert (1993) Topography of descending projections from the inferior colliculus to auditory brainstem nuclei in the rat. J. Comp. Neurol., *328:* 377–392.

Cajal, S.R. (1952) Histologie du Système Nerveux de l'Homme et des Vertébrés. Instituto Ramon y Cajal, Madrid.

Cant, N.B. (1992) The cochlear nucleus: Neuronal types and their synaptic organization. *In* The Mammalian Auditory Pathway: Neuroanatomy (ed. by D.B. Webster, A.N. Popper and R.R. Fay), Springer, New York, pp. 66–116.

Carr, C.E., and L. Maler (1986) Electroreception in gymnotiform fish: Central anatomy and physiology. *In* Electroreception (ed by T.H. Bullock and W. Heiligenberg), Wiley, New York, pp. 319–374.

Conley, R.A., and D. Bodznick (1994) The cerebellar dorsal granular ridge in an elasmobranch has proprioceptive and electroreceptive representations and projects homotopically to the medullary electrosensory nucleus. J. Comp. Physiol. A, *174:* 707–721.

Devor, A. (2000) Is the cerebellum like cerebellum-like structures? Brain Res. Rev., *34:* 149–156.

Eccles, J., M. Ito, and J. Szentagothai (1967) The Cerebellum as a Neuronal Machine. Springer, New York.

Finger, T.E., and S.L. Tong (1984) Central organization of eighth nerve and mechanosensory lateral line systems in the brainstem of ictalurid catfish. J. Comp. Neurol., *229:* 129–151.

Floris, A., M. Dino, D.B. Jacobowitz, and E. Mugnaini (1994) The unipolar brush cells of the rat cerebellar cortex and cochlear nucleus are calretinin-positive: A study by light and electron microscopic immunocytochemistry. Anat. Embryol., *189:* 495–520.

Fredette, B.J., and E. Mugnaini (1991) The GABAergic cerebello-olivary projection in the rat. Anat. Embryol., *184:* 225–243.

Fujino, K., and D. Oertel (2002) Bidirectional plasticity in the dorsal cochlear nucleus. Assoc. Res. Otolaryngol. Abstr. *31.*

Futuyma, D.J. (1998) Evolutionary Biology (3rd ed.). Sinauer, Sunderland, MA.

Gellman, R., A.E. Gibson, and J.C. Houk (1985) Inferior olivary neurons in the awake cat: Detection of contact and passive body displacement. J. Neurophysiol., *54:* 40–60.

Hillman, D.E. (1969) Neuronal organization of the cerebellar cortex in amphibia and reptilia. *In* Neurobiology of Cerebellar Evolution and Development (ed. by R. Llinas), American Medical Association, Chicago, pp. 279–325.

Hjelmstad, G.O., G. Parks, and D. Bodznick (1996) Motor corollary discharge activity and sensory responses related to ventilation in the skate vestibulolateral cerebellum: Implications for electrosensory processing. J. Exp. Biol., *199:* 673–681.

Ito, M. (1984) The Cerebellum and Neural Control. Raven Press, New York.

Ito, M. (2001) Cerebellar long-term depression: Characterization, signal transduction, and functional roles. Physiol. Rev., *81:* 1143–1195.

Johnson, J.I., J.A. Kirsch, R.L. Reep, and R.C. Switzer (1994) Phylogeny through brain traits: More characters for the analysis of mammalian evolution. Brain Behav. Evol., *43:* 319–347.

Kane, E.C. (1974) Synaptic organization in the dorsal cochlear nucleus of the cat: A light and electron microscopic study. J. Comp. Neurol., *155:* 301–330.

Kanold, P.O., and E.D. Young (2001) Proprioceptive information from the pinna provides somatosensory input to cat dorsal cochlear nucleus. J. Neurosci., *21:* 7848–7858.

Kotchabhakdi, N. (1976) Functional circuitry of the goldfish cerebellum. J. Comp. Physiol. A, *112:* 43–73.

Larsell, O. (1967) The Comparative Anatomy and Histology of the Cerebellum from Myxinoids Through Birds. University of Minnesota, Minneapolis, MN.

Lidierth, M., and R. Apps (1990) Gating in the spino-olivocerebellar pathways to the c1 zone of the cerebellar cortex during locomotion in the cat. J. Physiol., *430:* 453–469.

Liem, K., W. Bemis, W. Walker, and L. Grande (2001) Functional Anatomy of the Vertebrates: An Evolutionary Perspective. Harcourt, Philadelphia, PA.

Llinas, R., and C. Nicholson (1969) Electrophysiological analysis of alligator cerebellar cortex: A study of dendritic spikes. *In* Neurobiology of Cerebellar Evolution and Development (ed. by R. Llinas), American Medical Association, Chicago, IL, pp. 431–465.

Llinas, R., and Y. Yarom (1986) Oscillatory properties of guinea-pig inferior olivary neurones and their pharmacological modulation: an in vitro study. J. Physiol., *376:* 163–182.

Maeda, N., M. Niinobe, and K. Mikoshiba (1990) A cerebellar Purkinje cell marker P400 protein is an inositol 1,4,5-trisphosphate (InsP3) receptor protein. Purification and characterization of InsP3 receptor complex. EMBO J., *9:* 61–67.

Maekawa, K., and J.I. Simpson (1972) Climbing fiber activation of Purkinje cells in the flocculus by impulses transferred through the visual pathway. Brain Res., *39:* 245–251.

Maler, L., E.K.B. Sas, and J. Rogers (1981) The cytology of the posterior lateral line lobe of high frequency weakly electric fish (Gymnotidae): Dendritic differentiation and synaptic specificity in a simple cortex. J. Comp. Neurol., *195:* 87–140.

Martinez, S., P.H. Crossley, I. Cobos, J.L. Rubenstein, and G.R. Martin (1999) FGF8 induces formation of an ectopic isthmic organizer and isthmocerebellar development via a repressive effect on Otx2 expression. Development, *126:* 1189–1200.

McCormick, C.A. (1997) Organization and connections of octaval and lateral line centers in the medulla of a clupeid, *Dorosoma cepedianum.* Hearing Res., *110:* 39–60.

McCormick, C.A. (1999) Anatomy of the central auditory pathways of fish and amphibians. *In* Comparative Hearing: Fish and Amphibians (ed. by R.R. Fay and A.N. Popper), Springer, New York, pp. 155–217.

McCormick, C.A., and D.V. Hernandez (1996) Connections of octaval and lateral line nuclei of the medulla in the goldfish, including the cytoarchitecture of the secondary octaval population in goldfish and catfish. Brain Behav. Evol., *47:* 113–137.

Meek, J. (1992a) Why run parallel fibers parallel? Teleostean Purkinje cells as possible coincidence detectors, in a timing device subserving spatial coding of temporal differences. Neuroscience, *48:* 249–283.

Meek, J. (1992b) Comparative aspects of cerebellar organization. Eur. J. Morphol., *30:* 37–51.

Meek, J., and R. Nieuwenhuys (1991) Palisade pattern of mormyrid Purkinje cells: A correlated light and electron microscopic study. J. Comp. Neurol., *306:* 156–192.

Meek, J., and N.A.M. Schellart (1978) A Golgi study of goldfish optic tectum. J. Comp. Neurol., *182:* 89–122.

Meek, J., K. Grant, and C. Bell (1999) Structural organization of the mormyrid electrosensory lateral line lobe. J. Exp. Biol., *202:* 1291–1300.

Merzenich, M.M., L. Kitzes, and L. Aitkin (1973) Anatomical and physiological evidence for auditory specialization in the mountain beaver *(Aplodontia rufa).* Brain Res., *58:* 331–344.

Montgomery, J.C., and D. Bodznick (1993) Hindbrain circuitry mediating common mode suppression of ventilatory reafference in the electrosensory system of the little skate *Raja erinacea.* J. Exp. Biol., *183:* 203–215.

Montgomery, J.C., S. Coombs, R.A. Conley, and D. Bodznick (1995) Hindbrain sensory processing in lateral line, electrosensory, and auditory systems: A comparative overview of anatomical and functional similarities. Aud. Neurosci., *1:* 207–231.

Mortimer, J.A. (1975) Cerebellar responses to teleceptive stimuli in alert monkeys. Brain Res., *83:* 369–390.

Naciemiento, A.C. (1969) Spontaneous and evoked discharges of cerebellar Purkinje cells in the frog. *In* Neurobiology of Cerebellar Evolution and Development (ed. by R. Llinas), American Medical Association, Chicago, IL, pp. 373–395.

Nelson, M.E., and M.G. Paulin (1995) Neural simulations of adaptive reafference suppression in the elasmobranch electrosensory system. J. Comp. Physiol. A, *177:* 723–736.

Nieuwenhuys, R. (1967) Comparative anatomy of the cerebellum. Prog. Brain Res., *25:* 1–93.

Nieuwenhuys, R., H.J. Ten Donkelaar, and C. Nicholson (1997) The Central Nervous System of Vertebrates. Springer, Heidelberg.

Northmore, D.P.M., B. Williams, and H. Vanegas (1983) The teleostean torus longitudinalis: Responses related to eye movements, visuotopic mapping, and functional relations with the optic tectum. J. Comp. Physiol. A, *150:* 39–50.

Oertel, D., and S.H. Wu (1989) Morphology and physiology of cells in slice preparations of the dorsal cochlear nucleus of mice. J. Comp. Neurol., *283:* 228–247.

Paulin, M.G. (1993) The role of the cerebellum in motor control and perception. Brain Behav. Evol., *41:* 39–50.

Petralia, R.S., Y.X. Wang, H.M. Zhao, and R.J. Wenthold (1996) Ionotropic and metabotropic glutamate receptors show unique postsynaptic, presynaptic, and glial localizations in the dorsal cochlear nucleus. J. Comp. Neurol., *372:* 356–383.

Precht, W., J.I. Simpson, and R. Llinas (1976) Responses of Purkinje cells in rabbit nodulus and uvula to natural vestibular and visual stimuli. Pflügers Arch., *367:* 1–6.

Puzdrowski, R.L., and R.B. Leonard (1993) The octavolateral systems in the stingray, *Dasyatis sabina.* I. Primary projections of the octaval and lateral line nerves. J. Comp. Neurol., *332:* 21–37.

Roberts, P.D., and C.C. Bell (2000) Computational consequences of temporally asymmetric learning rules: II. Sensory image cancellation. J. Comput. Neurosci., *9:* 67–83.

Robertson, L.T. (1985) Somatosensory representation of the climbing fiber system in the rostral intermediate cerebellum. Exp. Brain Res., *61:* 73–86.

Ryugo, D.K., T. Pongstaporn, D.D. Wright, and A.H. Sharp (1995) Inositol 1,4,5-trisphosphate receptors: Immunocytochemical localization in the dorsal cochlear nucleus. J. Comp. Neurol., *358:* 102–118.

Saidel, W.M., and A.B. Butler (1997) Visual connections of the atypical diencephalic nucleus rostrolateralis in *Pantodon buchholzi* (Teleostei, Osteoglossomorpha). Cell Tissue Res., *287:* 91–99.

Schmidt, A.W., and D. Bodznick (1987) Afferent and efferent connections of the vestibulolateral cerebellum of the little skate, *Raja erinacea.* Brain Behav. Evol., *30:* 282–302.

Singer, W. (1995) Development and plasticity of cortical processing architectures. Science, *270:* 758–764.

Sotelo, C. (1969) Ultrastructural aspects of the cerebellar cortex of the frog. *In* Neurobiology of Cerebellar Evolution and Development (ed. by R. Llinas), American Medical Association, Chicago, IL, pp. 327–371.

Szabo, T., F. Haugede-Carre, and S. Libouban (1979) Cerebellar afferents in weakly electric mormyrid fish. Neurosci. Lett. Suppl., *3:* S144-S144.

Vanegas, H., B. Williams, and J.A. Freeman (1979) Responses to stimulation of marginal fibers in the teleostean optic tectum. Exp. Brain Res., *34:* 335–349.

Weedman, D.L., and D.K. Ryugo (1996) Projections from auditory cortex to the cochlear nucleus in rats: Synapses on granule cell dendrites. J. Comp. Neurol., *371:* 311–324.

Weinberg, R.J., and A. Rustioni (1987) A cuneocochlear pathway in the rat. Neuroscience, *20:* 209–219.

Wolff, A., and H. Kunzle (1997) Cortical and medullary somatosensory projections to the cochlear nuclear complex in the hedgehog tenrec. Neurosci. Lett., *221:* 125–128.

Wray, G.A. (2002) Do convergent developmental mechanisms underlie convergent phenotypes? Brain Behav. Evol., *59:* 327–336.

Wulliman, M.F. (1994) The teleostean torus longitudinalis: A short review on its structure, histochemistry, connectivity, possible function and phylogeny. Eur. J. Morphol., *32:* 235–242.

Brain, Behavior and Evolution

Brain Behav Evol 2002;59:327–336

Do Convergent Developmental Mechanisms Underlie Convergent Phenotypes?

Gregory A. Wray

Department of Biology, Duke University, Durham, N.C., USA

Key Words

Convergence · Homology · Evolution of development · Larva · *Hox* complex

Abstract

Convergence is a pervasive evolutionary process, affecting many aspects of phenotype and even genotype. Relatively little is known about convergence in developmental processes, however, nor about the degree to which convergence in development underlies convergence in anatomy. A switch in the ecology of sea urchins from feeding to nonfeeding larvae illustrates how convergence in development can be associated with convergence in anatomy. Comparisons to more distantly related taxa, however, suggest that this association may be limited to relatively close phylogenetic comparisons. Similarities in gene expression during development provide another window into the association between convergence in developmental processes and convergence in anatomy. Several well-studied transcription factors exhibit likely cases of convergent gene expression in distantly related animal phyla. Convergence in regulatory gene expression domains is probably more common than generally acknowledged, and can arise for several different reasons.

Introduction

Convergence pervades the biological world. From anatomy to protein sequences, it appears at all levels of biological organization; and from fungi to fishes, it permeates the evolutionary history of all groups of organisms [Darwin, 1859; Raff, 1996; Sanderson and Hufford, 1996; Wray, 1996; Moore and Willmer, 1997; Abouheif, 1999; Peintner et al., 2001; Eisthen, 2002; Nishikawa, 2002; Zakon, 2002]. The most celebrated cases of convergence, such as the eyes of cephalopods and vertebrates, are astonishingly similar in detail. Many other spectacular examples of convergence could be cited: bees attempt to mate with orchids, trouts are infected by parasitic clams, and birds vomit distasteful butterflies, all because of convergence.

How does convergence come about? Like many questions in biology, this is really two separate questions. The first issue is evolutionary and ecological: why convergence has evolved. The two classic categories of explanation are natural selection and phylogenetic constraint. By considering a situation in detail, it is often possible to provide more precise answers within each of these broad categories [Eisthen, 2002]. The second issue is genetic and developmental: how convergence has evolved. Here the questions concern the underlying basis for the similarity. The two categories of explanation are that homologous struc-

KARGER

Fax + 41 61 306 12 34
E-Mail karger@karger.ch
www.karger.com

Accessible online at:
www.karger.com/journals/bbe

Gregory A. Wray
Department of Biology
Duke University, Box 90338
Durham, NC 27708-0338 (USA)
Tel. +1 919 686 6696, Fax +1 919 660 7293, E-Mail gwray@duke.edu

tures, developmental processes, or genes were the substrate for the convergent feature, or that they were not. A complete understanding of any particular convergence can only come from answering both the 'why' and the 'how' questions [Raff, 1996].

This paper examines the relationship between convergence at one level of biological organization (regulatory genes or developmental mechanisms) with that at another (anatomy). Is anatomical convergence the result of conserved, convergent, or independent genetic and developmental bases? Addressing this deceptively simple question necessitates thinking about both the 'how' and the 'why' of convergence. Perhaps unsurprisingly, the answer is not simple. Convergence in anatomy is sometimes, but by no means always, due to convergent developmental processes; conversely, convergent developmental mechanisms sometimes, but not always, produce convergent structures. The same is true when we probe deeper, into the regulatory genes that control developmental processes.

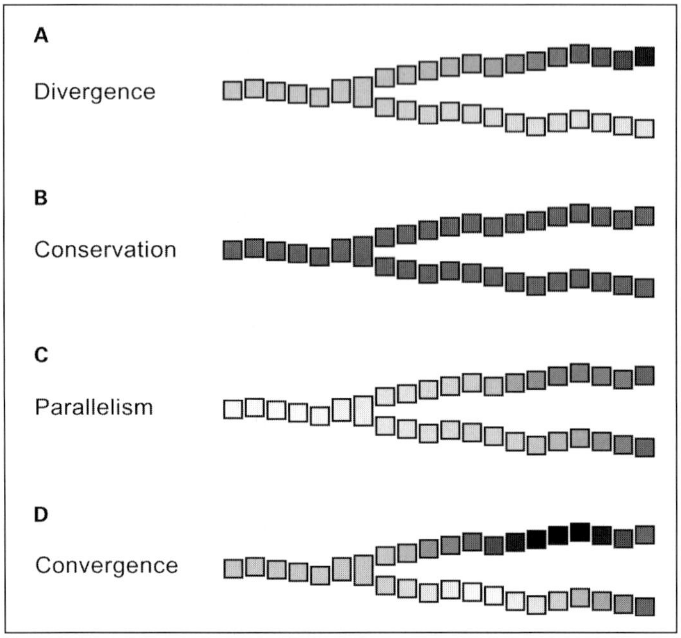

Fig. 1. Divergence (**A**), conservation (**B**), parallelism (**C**), and convergence (**D**). Similarities and differences in living organisms are observations, but divergence, conservation, parallelism, and convergence are all inferences about evolutionary history. As such, any interpretation of convergence needs to be explicitly tested against other possibilities. The four basic categories of evolutionary history are illustrated here. Time progresses from left to right. When two living organisms differ in some feature (boxes at tips of phylogeny, right-hand end), one or the other or both must differ in that regard from the latest common ancestor, an interpretation of divergence. When two extant species share a common feature, that similarity may have come about through various evolutionary histories. The simplest, and most commonly invoked, hypothesis is conservation. This hypothesis posits that the latest common ancestor possessed the same feature as its descendants. Particularly when the two extant species are distantly related, conservation cannot safely be assumed and other possibilities need to be explicitly tested. A hypothesis of parallelism posits that the feature was different in the latest common ancestor, and that it changed independently in both lineages to the same final state. A hypothesis of convergence is the most complex of the four basic possibilities: it posits that the feature was different in the latest common ancestor, and that it arrived at a common state in the two extant species by different routes.

Convergence and History

Convergence is not an observation – it is a specific hypothesis about evolutionary history [Brooks, 1996; Doyle, 1996; Abouheif, 1999]. What we have available for direct study are features of living species. For some taxa, we can also extract important (but rarely complete) anatomical information from the fossil record. For biochemical, genetic, developmental, physiological, and behavioral features, however, we are largely limited to species alive today.

The features of living organisms are a reflection of evolutionary history, but they are not always an easily interpreted guide to that history. Any similarity or difference that we observe in living organisms could have come about in a variety of ways [Darwin, 1859; Wake, 1999]. If two species differ in some feature of interest, we can confidently conclude that this is the result of divergence since they split from a common ancestor (fig. 1A). We cannot be confident, however, about what that feature was like in the latest common ancestor. It might have resembled the feature as manifest in one or the other living species, it might have been intermediate between them, or it might have been quite different from either.

If the feature of interest is similar among species, the situation is even more difficult to reconstruct – particularly if those species are not closely related. The simplest possibility is conservation: the feature has not changed since the two species last shared a common ancestor (fig. 1B). This is a common conclusion, and one that is often uncritically applied. There are two other basic possibilities, however. Parallelism occurs when a feature changes in the same way independently in two lineages (fig. 1C). The most complicated of the basic cases is convergence. This requires an initial period of divergence, followed by a period when the feature becomes more and more similar (fig. 1D). The essential feature of convergence is that similar endpoints are reached by different routes.

Distinguishing among conservation, parallelism, and convergence is a process of inference, not of direct observation, because we typically lack direct access to all but the terminal stages of evolutionary change. The obvious exception, mentioned earlier, is anatomy. Even for groups with a relatively good fossil record, however, distinguishing among conservation, parallelism, and convergence is far from straightforward [Sanderson and Hufford, 1996; Wake, 1999]. For features other than anatomy, identifying likely cases of convergence and ruling out other possible explanations for similarity is even more difficult [Wray, 1996; Abouheif, 1999; Bell, 2002; Eisthen, 2002]. Of course, ruling out convergence is equally difficult, as it requires demonstrating stronger support for an alternative interpretation.

The key to recovering biological history and interpreting it is a robust phylogenetic framework [Brooks and McLennan, 1991]. The reason we know that ichthyosaurs and cetaceans represent a spectacular case of convergence is that they are not close relatives within the tetrapod radiation. Conversely, the unobvious similarities that exist in the anatomy of hyraxes and elephants we interpret as conservation rather than convergence, because the two groups are very closely related. Although necessary, this reliance on phylogenetic relationship to recognize convergence introduces a bias: it is much easier to detect convergence when the species being compared are distantly related. Among closely related taxa, almost any similarity is routinely interpreted as conservation, but it would be surprising if some of these similarities did not turn out to be convergence.

Before moving on, a couple of additional points need to be emphasized. First, the basic categories of evolutionary pattern shown in figure 1 apply to any feature of an organism we might be interested in. Convergence is evident in protein sequences, development, physiology, anatomy, and, as mentioned elsewhere in this issue, in the structure and function of the nervous system [Wray, 1996; Messler and Stewart, 1997; Bell, 2002; Eisthen, 2002; Nishikawa, 2002; Zakon, 2002]. Second, the scenarios shown in figure 1 are the simplest possible, idealized cases. In real organisms, the features we are most interested in often have complex evolutionary histories made up of combinations of these processes [Brooks, 1996; Doyle, 1996; Abouheif, 1999; Bell, 2002]. And third, different features can have different evolutionary histories within the same group of organisms. An animal may contain some features that are plesiomorphic (ancestral) and other features that are quite specialized. A platypus is 'primitive' for a mammal in that it lays eggs, but also possesses a sophisticated electrosensory system for detecting prey that is highly specialized.

Convergence in Development and Anatomy

In order to understand how and why a particular case of convergence has evolved, it is useful to examine the relationship between convergent phenotypes and the developmental processes that generate them. The convergent evolution of larval form in sea urchins and other marine invertebrates illustrates the kind of information one can draw from such a comparison when it is made across multiple taxa.

Sea urchins have a complex life history, with most species developing by means of a planktotrophic (feeding) larva that bears little resemblance to the adult it develops into (fig. 2). Much of the anatomy of planktotrophic larvae is geared towards grazing efficiently on single-celled algae, including a functional digestive tract and a skeleton that supports arms that increase feeding rates [Strathmann, 1971; Hart, 1991]. These features are widely distributed among the living groups of sea urchins and represent the ancestral state for the group [Strathmann, 1985; Wray, 1992]. Based on the phylogenetic distribution of traits and the fossil record, it is likely that a core set of anatomical features has been conserved for at least 250 million years, just before the major lineages of living sea urchins radiated [Wray, 1992].

Although widespread and ancient, the presence of planktotrophic larvae in the life cycle is not universal among sea urchins. About 20% of living species instead undergo an abbreviated mode of development [Emlet et al., 1987; Raff, 1987; Emlet, 1990]. In most cases, a larva is still present in the life cycle, but it is lecithotrophic (deriving nutrition from yolk). Phylogenetic evidence indicates that lecithotrophic larvae have evolved from planktotrophic larvae on at least a dozen separate occasions [Strathmann, 1985; Raff, 1987; Wray, 1996; Nielsen, 1998]. These independently evolved lecithotrophic larvae are anatomically convergent: their overall form is streamlined, a mouth and functional digestive system are absent, the skeleton is reduced or absent, and the ciliated band is repositioned or lost [Emlet, 1991; Wray and Bely, 1994]. The degree of anatomical convergence is so striking that phylogenetic analysis based on larval anatomy group lecithotrophic species into a well-supported 'clade', even

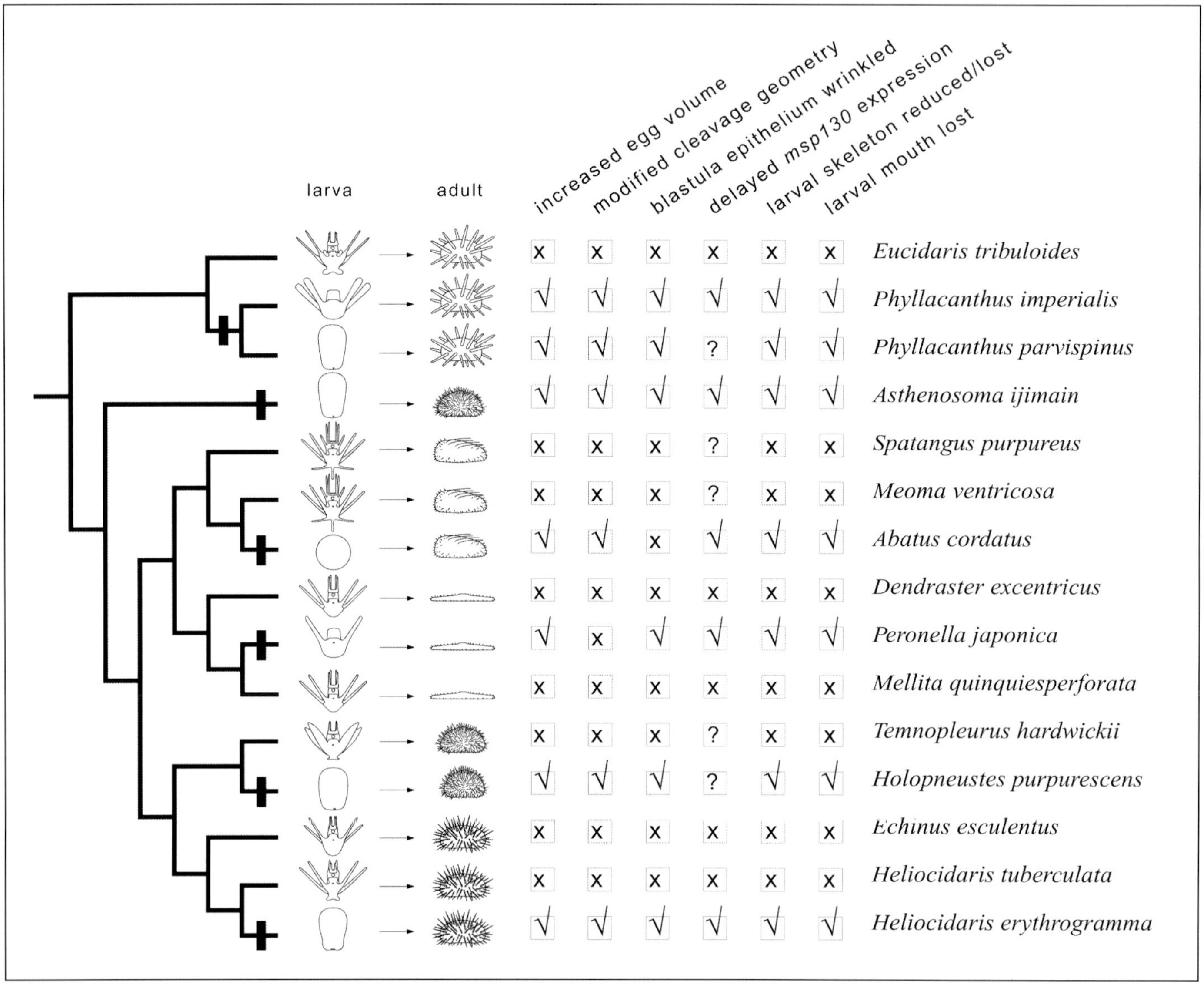

Fig. 2. Convergent evolution of embryonic development in echinoderms. The anatomy of sea urchin larvae shows significant discordance with phylogenetic history. The reconstructed ancestral state is a feeding larva with long arms and a complex internal skeleton [Strathmann, 1988; Wray, 1992]. Although these feeding larvae share many general traits, they have diverged in their detailed anatomy [Wray, 1992], as shown diagrammatically by the small cartoons. On several separate occasions, larval feeding has been lost (black bars on phylogeny) [Raff, 1987; Emlet et al., 1997]. In each case, larval anatomy is both extensively modified and convergently similar [Wray, 1996]. In each of these cases, underlying features of development have also changed in parallel or convergently: egg size [Emlet et al., 1987], egg composition [Byrne et al., 1999; Villinski et al., 2002], gemometry of early cell divisions [Wray and Bely, 1994], and timing of gene expression [Wray and Bely, 1994]. Thus, the anatomy of nonfeeding larvae in sea urchins provides a relatively well documented case of an association between convergent evolution in anatomy and convergent evolution in development. [Phylogeny from Smith et al., 1995.]

though both adult anatomy and sequence data support many independent origins [Wray, 1996].

Underlying anatomical convergences in lecithotrophic larvae are convergences in several developmental processes. The eggs of these species are invariably much larger, as might be expected, given that the larvae rely on maternal provisions rather than feeding to sustain them until metamorphosis [Emlet et al., 1987; Raff, 1987]. The geometry of early cell divisions in the embryo are modified in almost all cases, with a shift from unequal vegetal-most fourth and fifth cleavages to equal cleavages throughout the early embryo [Wray and Bely, 1994]. That this change has evolved repeatedly is rather remarkable, as the micromere daughters of the unequal fourth cleavage operate as the 'organizer' region of the early embryo in species with planktotrophic larvae [Davidson, 2001]. Additional convergently similar changes in early development include a characteristic transient wrinkling of the epithelium during the blastula stage, higher cell numbers during gastrulation, earlier morphogenesis of the coeloms, and reduced time to metamorphosis [Wray and Bely, 1994; Wray, 1996].

Gene expression during oogenesis and in embryos is also convergently similar in species with lecithotrophic larvae. The eggs of lecithotrophs are maternally provisioned with proportionally less vitellogenin and proportionally more wax esters than those of planktotrophs [Byrne et al., 1999; Villinski et al., 2002]. Because wax esters provide a rich source of energy, it makes sense that they would be loaded into the eggs of lecithotrophic larvae, but why this particular compound has been repeatedly utilized and not some other energy-rich molecule is unclear. The parallel proportional reduction in vitellogenin content is also mysterious, as vitellogenin is yolk protein and might be expected to be loaded at higher concentrations in the eggs of lecithotrophs. The fact that vitellogenin is metabolized primarily after metamorphosis might mean that it is functionally tied to larval nutritional mode only indirectly. Zygotic expression of the gene *msp130* is also convergently similar in species with lecithotrophic larvae. This gene encodes a structural glycoprotein that is deposited in and around the biomineral matrix of the larval skeleton [Leaf et al., 1987]. In species with planktotrophic larvae, *msp130* is first expressed during the early phases of gastrulation, which is just before skeletogenesis begins [Wray and McClay, 1989]. Expression is delayed until after gastrulation in lecithotrophs, whose larval skeletons are either lacking or greatly reduced [Parks et al., 1988, 1989; Amemiya and Emlet, 1992]. This has presumably freed them from the need to activate early zygotic *msp130* expression, and indeed the onset of expression is more commonly associated with the appearance of adult skeleton during metamorphosis in lecithotrophs [Wray and Bely, 1994].

Lecithotrophic larvae have also evolved from planktotrophic larvae in other groups of marine invertebrates. Examples include other echinoderm classes, enteropneust hemichordates, nemerteans, sipunculids, and polychaete annelids [Jägersten, 1972; Strathmann, 1985; Nielsen, 1998]. The planktotrophic larvae of these different phyla and classes are anatomically distinctive, with characteristic features that reflect different mechanisms of feeding, locomotion, and defense [Emlet, 1991; Nielsen, 1998]. The lecithotrophic larvae that have evolved in each of these clades do not resemble closely related planktotrophic larvae; instead, they are convergently similar to lecithotrophic larvae in other classes and phyla [Emlet, 1991].

The reasons why lecithotrophic larvae look similar are probably tied to the parallel ecological situations that they experience [Emlet, 1991; Wray, 1996]. Planktotrophic larvae are not competent to metamorphose initially, and must feed for days or weeks before they have put on enough weight to complete metamorphosis. Few individuals survive this planktonic phase of the life cycle, in which mortality rates are probably on the order of 10–20% per day [Morgan, 1995]. Thus, planktotrophic larvae experience strong selection for efficient feeding and resistance to predation [Havenhand, 1995]. Lecithotrophic larvae, in contrast, do not need to feed and are probably under strong selection to shorten this hazardous phase of the life cycle [Hart, 1996]. Structures that are primarily involved in feeding, such as larval skeletons and functional guts, are reduced or lost [Wray, 1996]. Ciliated bands, which are used for both feeding and propulsion in planktotrophs, are repositioned to body edges and perpendicular to the anteroposterior axis so as to maximize propulsion in lecithotrophs [Emlet, 1991]. The result is typically a larva with an ovoid body shape, no mouth, and parallel ciliated bands. These phylogenetically unrelated but ecologically and anatomically similar larvae are called schmoos [Wray, 1996].

Whereas early development and gene expression are convergently similar in sea urchins with lecithotrophic larvae, this is not the case when comparisons are extended to more distantly related taxa. The reasons are probably tied to phylogenetic constraints. Although lecithotrophy has evolved independently in several different families and orders of sea urchins [Emlet et al., 1987; Raff, 1987], these clades all share many similarities in embryonic and larval development; in contrast, comparisons across phyla involve cases where lecithotrophy evolved in species with far fewer similarities in mechanisms of early development [Strathmann, 1985; Nielsen, 1998]. Given parallel ecological conditions favoring similar changes in larval anatomy, the odds of producing a particular anatomical change in the same way are likely to be higher in more closely related species that share a greater proportion of their developmental mechanisms. Conversely, for species that share few features of development of planktotrophic larvae, the evolution of lecithotrophy is more likely to involve different genetic and developmental bases; yet the similar ecological conditions will favor convergently similar overall phenotypes.

This extended example illustrates two general points about development and convergence. First, similar ecological conditions can produce remarkably convergent developmental modes. This should not be surprising, given the many remarkable cases of convergence in anatomy and physiology that have been documented. Few cases of convergence in development have been studied in detail, but they may not be as unusual as the small number of published examples would suggest. Second, convergence in development is probably inversely related to phylogenetic distance. In relatively closely related species, convergent anatomies may be produced by convergent developmental and molecular processes. In more distantly related species, the likelihood increases that different developmental and molecular mechanisms will contribute to a convergently similar anatomy. This is simply because the number of differences in developmental mechanisms will increase with phylogenetic distance. Thus, convergence at one level of biological organization (anatomy) is not always associated with convergence at another (development), but we can sometimes predict when they will and will not be associated.

Convergence in Developmental Gene Expression

A particularly interesting issue emerging from the field of 'evo-devo' is convergent evolution of gene expression and function during development [Abouheif et al., 1998; Tabin et al., 1999; Wray and Lowe, 2000]. How common is convergence in developmental regulatory gene expression? How similar do such cases appear? There is a widespread, although generally tacit, assumption that convergent gene expression is highly unlikely. This assumption is implicit, for example, when using comparisons of developmental gene expression to make inferences about the homology of structures [e.g., Finkelstein and Boncinelli, 1994; DeRobertis and Sasai, 1996; Holland and Holland, 1999; Lee and Jacobs, 1999]. Yet no serious attempt has been made to examine this issue directly; nor is there any obvious theoretical basis to support the assumption that convergence in gene expression should be rare. Indeed, several likely cases of convergence in regulatory gene expression have now been documented. The implications are quite interesting, both for the exercise of identifying homologous structures in distantly related organisms, and for understanding how modifications in gene expression lead to the evolution of phenotype.

A few cases of convergently similar gene expression were discussed in the previous section. The delay in *msp130* expression that has accompanied the evolution of lecithotrophy in echinoids on several occasions is perhaps the clearest. Convergent changes in gene expression are also implied by what we know of egg composition in lecithotrophs. In particular, the switch to wax esters probably required convergent changes in the expression of the enzymes that produce these compounds. Nor should we expect such cases to be particularly rare. Natural populations may harbor considerable genetic variation for differences in the expression of genes encoding enzymes and other structural proteins [Segal et al., 1999; Rockman and Wray, in press]. If this is the case, then natural selection should be able to sort this variation in similar ways in similar environments, leading to convergent gene expression.

Regulatory genes are of even more interest, because changes in their expression and function are widely considered to be involved in the evolution of morphology [Wilson, 1975; Raff and Kaufman, 1983; Raff 1996; Carroll et al., 2001]. These are also the genes most commonly used to make inferences about homology of structures among distantly related animals [Abouheif et al., 1997; Holland and Holland, 1999]. The premiere example of regulatory genes are members of the *Hox* complex that encode homeodomain transcription factors. The *Hox* genes famously specify position along the anteroposterior axis in *Drosophila melanogaster, Tribolium castaneum, Mus musculus*, and *Caenorhabditis elegans*, the only species for which strong functional data exist [Gerhart and Kirschner, 1997; Carroll et al., 2001; Davidson, 2001; Wilkins, 2002]. These genes are expressed in nested domains along the anteroposterior axes of many more animals, and are generally assumed to play the same anteroposterior patterning role throughout at least the Bilateria [DeRobertis and Sasai, 1996; Carroll et al., 2001].

An equally remarkable point, but one that has received far less comment, is that the *Hox* complex is expressed in many non-homologous structures in different phyla. For example, in insects its first expression domain encompasses almost all cells within the early embryo, and later it is expressed in the gut; in mice it is first expressed within the hindbrain, and later within the somites, limb buds, and reproductive tract; and in sea urchins, it is expressed in the coeloms [reviewed in: Duboule, 1995; Carroll et al., 2001; Davidson, 2001]. Few of these sites of expression are homologous. Some expression domains encompass

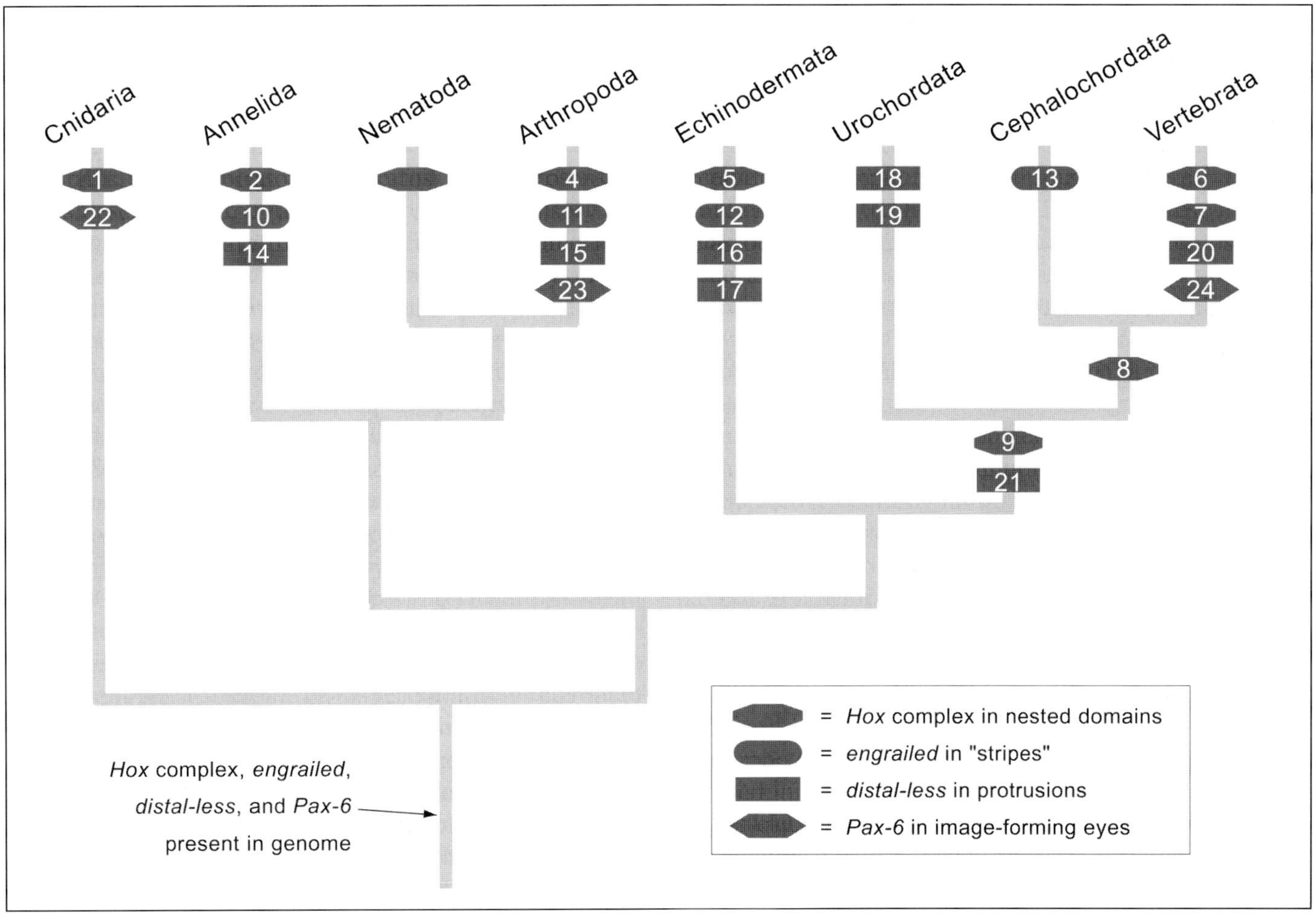

Fig. 3. Convergent evolution of regulatory gene expression domains. Similarities in gene expression are almost always interpreted as conserved, but it has become clear that some similarities in gene expression are more plausibly interpreted as cases of convergence. Some likely cases of convergence in the expression of four gene families are illustrated here for metazoans. *Hox* complex: 1 planula/polyp ectoderm, 2 central nervous system, 3 various scattered cells, 4 embryonic blastoderm, 5 larval coeloms, 6 fin/limb buds, 7 reproductive tract, 8 hindbrain, 9 somites. *Engrailed:* 10 ectoderm in leeches (but not more basal polychaetes), 11 embryonic blastoderm, 12 ectoderm of adult arm in ophiuroids (but not other echinoderms), 13 somites. *Distal-less:* 14 ectoderm of parapodia, 15 gills/appendages, 16 podia (tubefeet), 17 brachiolar arms of asteroid larvae (but not in other echinoderm larvae), 18 ampullae of larvae, 19 siphons of adults, 20 fin/limb buds, 21 gill bars. *Pax-6:* 22 rhopalia of medusae, 23 compound eyes, 24 eyes. See text for discussion and references.

almost all of the body and others only a small fraction of it, whereas various sites of expression are distributed across all three germ layers and a surprising variety of organs. Homology is particularly implausible among different expression domains within the same species (e.g., hindbrain and leg). The inevitable conclusion is that the expression of the *Hox* complex has evolved into new locations and times during development on several separate occasions (fig. 3) [Duboule and Wilkins, 1998; Wray and Lowe, 2000; Davidson, 2001; Wilkins, 2002].

Furthermore, many of these *Hox* redeployments are convergently similar. Although clearly independently evolved, each of these cases exhibits colinearity, a striking and detailed similarity in gene expression. Colinearity means that the spatial order of each *Hox* paralogue's expression within the embryo corresponds to its location along the chromosome. This detailed similarity of expression is not surprising at one level. Transcriptional regulation within the *Hox* complex is thought to involve a molecular 'bucket brigade', such that the most 5′ ('anterior') paralogue can activate the paralogue 3′ to it, which in turn can activate the next paralogue, and so forth [Gerhart and Kirschner, 1997; Davidson, 2001]. Thus, any mutation that activates transcription of a *Hox* gene in a new location might also indirectly activate all the genes 3′ to it within the complex.

The *Hox* complex is a spectacular case of molecular convergence, but hardly the only example. The *engrailed* gene, which also encodes a homeodomain transcription factor, is required to establish segment boundaries in *Drosophila*. It is expressed in segmentally-repeated stripes along the anteroposterior axis. In at least three other phyla, engrailed is also expressed in 'stripes', although in one case these stripes are mesodermal rather than ectodermal (fig. 3) [Wedeen and Wiseblat, 1991; Holland et al., 1996; Lowe and Wray, 1997]. Here the similarities are probably superficial rather than indicative of homology among the structures where the gene is expressed, because other genes involved in insect segmentation do not show similar stripes of expression in these phyla. In addition, in all three cases besides arthropods, other species within the same phylum lack engrailed 'stripes'. Of course, any gene that is expressed in an iterative structure such as ganglia or somites will itself display an iterative expression domain, potentially leading to convergent similarity that is superficial rather than indicative of common ancestry. This is probably one of the most common ways in which superficially similar patterns of gene expression arise.

A more complicated situation involves the *distal-less* gene (*dll* in insects, *Dlx* elsewhere), which also encodes a homeodomain transcription factor, but is not part of a gene complex. It is required for correct limb outgrowth in *Drosophila* [reviewed in Tabin et al., 1999]. Remarkably, *distal-less* is also expressed in the developing limbs and fins of vertebrates, the podia of sea urchins, the siphons and ampullae of ascidians, and parapodia of annelids (fig. 3) [Lowe and Wray, 1997; Panganiban et al., 1997]. *Distal-less* is thus expressed in several appendages that are almost certainly not homologous. Yet the spatial and temporal similarities in gene expression are surprisingly close: in nearly all cases, the gene is only expressed in the most distal portion of the appendage and primarily during appendage outgrowth.

There are two basic interpretations of these data: either the appendages really are homologous or *distal-less* shows a tendency towards convergent expression in appendages [Tabin et al., 1999; Wray and Lowe, 2000]. The first hypothesis is not as parsimonious at it first appears, as it requires dozens of losses of limbs (early vertebrates clearly lacked limbs, as did early ecdysozoans) and an extraordinary anatomical diversification of the putative ancestral limb (there are virtually no anatomical similarities between echinoderm podia and arthropod legs, for example). On the other hand, the second hypothesis seems to require a fantastic degree of coincidence. A possible solution is that *distal-less* is part of a regulatory pathway that, once activated by a key regulator, deploys itself in the same topological arrangement irrespective of where that activation takes place [Tabin et al., 1999; Wray and Lowe, 2000]. According to this view, the only difference from the *Hox* complex is that we don't know what the other members of the coordinately-deployed pathway are.

A final example involves *Pax-6/eyeless*, which encodes a paired-box transcription factor. This gene is famously expressed in the eyes of insects, vertebrates, mollusks, and nemerteans [Quiring et al., 1994; reviewed by Gehring and Ikeo, 1999]. Eyes, at least as image-forming organs, are clearly not homologous on anatomical grounds among these phyla; yet in the case of both insects and vertebrates, *Pax-6* is required for eye development. Once again, the similarities in gene expression are interpreted as convergently similar (fig. 3). The probable cause in this case is that these eyes, despite their disparate anatomical designs, all function using homologous photoreceptor cells. In other words, a component cell type is homologous, but the organs within which it is housed in different phyla are not. The situation is more complicated than this, however, because *Pax-6* does more than just specify the photoreceptor cell type, at least in *Drosophila* and *Mus*. In these animals it clearly plays a more fundamental role in regulating other aspects of eye development. Another complication is that *Pax-6* is expressed in many places other than eyes [Quiring et al., 1994; Wilkins, 2002]. Several specific hypotheses have been proposed to explain how the association with eyes could have evolved from an initial role in photoreceptor cell fate specification [Gehring and Ikeo, 1999; Davidson, 2001; Wilkins, 2002]. These hypotheses differ in detail, but they all present scenarios in which *Pax-6* has taken on additional roles in eye development from a starting point restricted to photoreceptor cell specification.

To summarize, several cases are known where the expression domains of key developmental regulatory proteins are convergently similar. Some cases of convergent similarity are probably the result of similar ecological circumstances. The structural genes discussed near the beginning of this section provide likely examples. However, many other cases of convergent similarity involve homologous genes that are expressed in a similar manner in structures that are almost certainly not homologous. Such cases are probably not particularly rare. The examples discussed above are among the most thoroughly studied, but by no means the only examples that could be cited. Superficial similarity in gene expression can probably arise in several ways: linked transcriptional regulation (genes of the *Hox* complex and *distal-less*), expression in meristic structure *(engrailed)*, and expression within a homologous cell type that is later incorporated in non-homologous organs *(Pax-6)*. This is probably not an exhaustive list of the possible causes of convergently similar gene expression.

Acknowledgments

Thanks to Heather Eisthen and Kiisa Nishikawa for inviting me to participate in the Karger Workshop, and for many stimulating discussions about development and evolution. Thanks also to Dr. Thomas Karger for supporting the workshop. My research is funded by the National Science Foundation and the National Aeronautics and Space Administration.

References

Abouheif, E. (1999) Establishing homology criteria for regulatory gene networks: Prospects and challenges: *In* Homology (ed. by G.R. Bock and G. Cardew), Wiley, Chichester, pp. 207–221.

Abouheif, E.M., M. Akam, W.J. Dickinson, P.W.H. Holland, A. Meyer, N.H. Patel, R.A. Raff, V.L. Roth, and G.A. Wray (1997) Homology and developmental genes. Trends Genet., *13:* 432–433.

Amemiya, S., R.B. Emlet (1992) The development and larval form of an echinothurioid echinoid, *Asthenosoma ijimai*, revisited. Biol. Bull., *182:* 15–30.

Bell, C.C. (2002) Evolution of cerebellum-like structures. Brain Behav. Evol., *59:* 312–326.

Brooks, D.R. (1996) Explanations of homoplasy at different levels of biological organization. *In* Homoplasy: The recurrence of Similarity in Evolution (ed. by M.J. Sanderson and L. Hufford), Academic Press, San Diego, pp. 3–36.

Brooks, D.R., D.A. McLennan (1991) Phylogeny, Ecology, and Behavior: A Research Program in Comparative Biology. Univ. of Chicago Press, Chicalo, IL.

Byrne, M., J.T. Villinski, P. Cisternas, R.K. Siegel, E. Popodi, and R.A. Raff (1999) Maternal factors and the evolution of developmental mode: Evolution of oogenesis in *Heliocidaris erythrogramma*. Dev. Genes. Evol., *209:* 275–283.

Carroll, S.B., J.K. Grenier, and S.D. Weatherbee (2001) From DNA to Diversity: Molecular Genetics and the Evolution of Animal Design. Blackwell Scientific, Malden, MA.

Darwin, C. (1859) On the Origin of Species. John Murray, London.

Davidson, E.H. (2001) Genomic Regulatory Systems: Evolution and Development. Academic Press, San Diego, CA.

DeRobertis, E.M., Y. Sasai (1996) A common plan for dorsoventral patterning in Bilateria. Nature, *380:* 37–40.

Doyle, J.J. (1996) Homoplasy connections and disconnections: Genes and species, molecules and morphology. *In* Homoplasy: The Recurrence of Similarity in Evolution (ed. by M.J. Sanderson and L. Hufford), Academic Press, San Diego, CA, pp 37–66.

Duboule, D. (ed) (1995) Guide to the Homeobox Genes. Oxford University Press, Oxford, UK.

Duboule D., and A.S. Wilkins (1998) The evolution of 'bricolage'. Trends Genet., *14:* 54–59.

Eisthen, H.L. (2002) Why are olfactory systems of different animals so similar? Brain Behav. Evol., *59:* 273–293.

Emlet, R.B. (1990) World patterns of developmental mode in echinoid echinoderms. *In* Advances in Invertebrate Reproduction (ed by M. Hoshi and O Yamashita), Elsevier, Amsterdam, pp 329–335.

Emlet, R.B. (1991) Functional constraints on the evolution of larval forms of marine invertebrates: Experimental and comparative evidence. Am. Zool., *31:* 707–725.

Emlet, E.B., L.R. McEdward, and R.R. Strathmann (1987) Echinoderm larval ecology viewed from the egg. *In* Echinoderm Studies (ed. by M. Jangoux and J.M. Lawrence), Balkema Press, Amsterdam, pp. 55–136.

Finkelstein, R., and E. Boncinelli (1994) From fly head to mammalian forebrain: The story of *otd* and *Otx*. Trends Genet., *10:* 310–315.

Gehring, W.J., and K. Ikeo (1999) *Pax6:* Mastering eye morphogenesis and eye evolution. Trends Genet., *15:* 371–377.

Gerhart, J., and M. Kirschner (1997) Cells, Embryos and Evolution. Blackwell Science, Malden, MA.

Hart, M.W. (1991) Particle capture and the method of suspension feeding by echinoderm larvae. Biol. Bull., *180:* 12–27.

Hart, M.W. (1996) Evolutionary loss of larval feeding: Development, form, and function in a facultatively feeding larva, *Brisaster latifrons*. Evolution, *50:* 174–187.

Havenhand, J. (1995) Evolutionary ecology of life histories. *In* Ecology of Marine Invertebrate Larvae (ed. by L.R. McEdward), CRC Press, Boca Raton, FL, pp 79–122.

Holland, N.D., and L.Z. Holland (1999) Amphioxus and the utility of molecular genetic data for hypothesizing body part homologies between distantly related animals. Am. Zool., *39:* 630–640.

Holland, N.D., G. Panganiban, E.L. Henyey, and L.Z. Holland (1996) Sequence and developmental expression of *AmphiDII*, an amphioxus *Distal-less* gene transcribed in the ectoderm, epidermis and nervous system: Insights into evolution of craniate forebrain and neural crest. Development, *122:* 2911–2920.

Jägersten, G. (1972) Evolution of the Metazoan Life Cycle. Academic Press, London, UK.

Leaf, D.S., J.A. Anstrom, J.-E. Chin, M.A. Harkey, and R.A. Raff (1987) Antibodies specific to a fusion protein identify a cDNA clone encoding *msp130*, a primary mesenchyme-specific cell surface protein of the sea urchin embryo. Dev. Biol., *121:* 29–40.

Lee S.E., and D.K. Jacobs (1999) Expression of Distal-less in molluscan eggs, embryos, and larvae. Evol. Dev., *1:* 172–179.

Lowe, C.J., and G.A. Wray (1997) Radical alterations in the roles of homeobox genes during echinoderm evolution. Nature, *389:* 718–721.

Messler, W. and C.B. Stewart (1997) Episodic adaptive evolution of primate lysozymes. Nature, *385:* 151–154.

Moore, J., and P. Willmer (1997) Convergent evolution in invertebrates. Biol. Rev., *72:* 1–60.

Morgan, S.G. (1995) Life and death in the plankton. *In* Ecology of Marine Invertebrate Larvae (ed. by L.R. McEdward), CRC Press, Boca Raton, FL, pp 279–321.

Nielsen, C. (1998) Origin and evolution of animal life cycles. Biol. Rev., *73:* 125–155.

Nishikawa, K. (2002) Evolutionary convergence in nervous systems: Insights from comparative phylogenetic studies. Brain Behav. Evol., *59:* 240–249.

Panganiban, G., S.M. Irvine, C.J. Lowe, H. Roehl, L.S. Corley, B. Sherbon, J.K. Grenier, J.F. Fallon, J. Kimble, M. Walker, G.A. Wray, B.J. Swalla, M.Q. Martindale, and S.B. Carroll (1997) The origin and evolution of animal appendages. Proc. Natl. Acad. Sci. USA, *94:* 5162–5166.

Parks, A.L., B.W. Bisgrove, G.A. Wray, and R.A. Raff (1989) Direct development in the sea urchin *Phyllacanthus parvispinus* (Cidaroidea): Phylogenetic history and functional modification. Biol. Bull., *177:* 96–109.

Parks, A.L., B.A. Parr, J.-E. Chin, D.S. Leaf, and R.A. Raff (1988) Molecular analysis of heterochronic changes in the evolution of direct developing sea urchins. J. Evol. Biol., *1:* 27–44.

Peintner, U., N.L. Bougher, M.A. Castellano, J.M. Moncalvo, M.M. Moser, J.M. Trappe, R. Vilgalys (2001) Multiple origins of sequestrate fungi related to *Cortinarius* (Cortinariaceae). Am. J. Bot., *88:* 2168–2179.

Quiring, R., U. Walldorf, U. Kloter, and W.J. Gehring (1994) Homology of the *eyeless* gene of *Drosophila* to the *Small eye* gene in mice and *Aniridia* in humans. Science, *265:* 785–789.

Raff, R.A. (1987) Constraint, flexibility, and phylogenetic history in the evolution of direct development in sea urchins. Dev. Biol., *119:* 7–19.
Raff, R.A. (1996) The Shape of Life. University of Chicago Press, Chicago, IL.
Raff, R.A., and T.C. Kaufman (1983) Genes, Development, and Evolution. McMillan, New York.
Rockman, M.V., and G.A. Wray (2002) Abundant raw material for *cis*-regulatory evolution in humans. Mol. Biol. Evol., in press.
Sanderson, M.J., and L. Hufford (eds.) (1996) Homoplasy: The Recurrence of Similarity in Evolution. Academic Press, San Diego, CA.
Segal, J.A., J.L. Barnett, and D.L. Crawford (1999) Functional analyses of natural variation in SP1 binding sites of a TATA-less promoter. J. Mol. Evol., *49:* 736–749.
Smith, A.B., D.T.J. Littlewood, and G.A. Wray (1995) Comparing patterns of evolution: Larval and adult life history stages and small subunit ribosomal RNA of post-Paleozoic echinoids. Phil. Trans. R. Soc. Lond. B, *349:* 11–18.
Strathmann, R.R. (1971) The feeding behavior of planktotrophic echinoderm larvae: Mechanisms, regulation and rates of suspension feeding. J. Exp. Mar. Biol. Ecol., *6:* 109–160.
Strathmann, R. (1985) Feeding and nonfeeding larval development and life-history evolution in marine invertebrates. Ann. Rev. Ecol. Syst., *16:* 339–361.
Strathmann, R.R. (1988) Functional requirements and the evolution of developmental patterns. *In* Echinoderm Biology: Proceedings of the Sixth International Echinoderm Conference (ed. by R.D. Burke, P.V. Mladenov, P. Lambert and R.L. Parsley), Balkema, Amsterdam, pp. 55–61.
Tabin, C.J., S.B. Carroll, and G. Panganiban (1999) Out on a limb: Parallels in vertebrate and invertebrate limb patterning and the origin of appendages. Am. Zool., *39:* 650–663.
Villinski, J.T., J.C. Villinski, M. Byrne, and R.A. Raff (2002) Convergent maternal provisioning and life history evolution in echinoderms. Evolution, in review.
Wake, D. (1999) Homoplasy, homology and the problem of 'sameness' in biology. *In* Homology (ed. by G.R. Bock and G. Cardew), Wiley, Chichester, pp. 24–33.
Wedeen, C.J., and D.A. Weisblat (1991) Segmental expression of an *engrailed*-class gene during early development and neurogenesis in an annelid. Development, *113:* 805–814.
Wilkins, A.S. (2002) The Evolution of Developmental Pathways. Sinauer, Sunderland, MA.
Wilson, A.C. (1975) Evolutionary importance of gene regulation. Stadler Symp., *7:* 117–134.
Wray, G.A. (1992) The evolution of larval morphology during the post-Paleozoic radiation of echinoids. Paleobiology, *18:* 258–287.
Wray, G.A. (1996) Parallel evolution of nonfeeding larvae in echinoids. Sys. Biol., *45:* 308–322.
Wray, G.A., A.E. Bely (1994) The evolution of echinoderm development is driven by several distinct factors. Development Suppl., 97–106.
Wray, G.A., and C.J. Lowe (2000) Developmental regulatory genes and echinoderm evolution. Sys. Biol., *49:* 28–51.
Wray, G.A., and D.R. McClay (1989) Molecular heterochronies and heterotopies in early echinoid development. Evolution, *43:* 803–813.
Zakon, H. (2002) Convergent evolution on the molecular level. Brain Behav. Evol., *59:* 250–261.

Subject Index Vol. 59, No. 5–6, 2002

Author Index Vol. 59, 2002

Brain, Behavior and Evolution

KARGER

Fax + 41 61 306 12 34
E-Mail karger@karger.ch
www.karger.com

Accessible online at:
www.karger.com/journals/bbe

Subject Index Vol. 59, 2002

Brain, Behavior
and Evolution

KARGER
Fax +41 61 306 12 34
E-Mail karger@karger.ch
www.karger.com

Accessible online at:
www.karger.com/journals/bbe

Vol. 59, 2002

Official Journal of the J.B. Johnston Club

Brain, Behavior and Evolution

Founder and editor W. Riss (1968–1986), continued by R.G. Northcutt (1986–1998)

KARGER

Basel · Freiburg · Paris · London · New York ·
New Delhi · Bangkok · Singapore · Tokyo · Sydney

S. Karger
Medical and Scientific Publishers
Basel · Freiburg · Paris · London
New York · New Delhi · Bangkok
Singapore · Tokyo · Sydney

Drug Dosage
The authors and the publisher have exerted every effort to ensure that drug selection and dosage set forth in this text are in accord with current recommendations and practice at the time of publication. However, in view of ongoing research, changes in government regulations, and the constant flow of information relating to drug therapy and drug reactions, the reader is urged to check the package insert for each drug for any change in indications and dosage and for added warnings and precautions. This is particularly important when the recommended agent is a new and/or infrequently employed drug.

Printed in Switzerland on acid-free paper by
Reinhardt Druck, Basel

KARGER

Fax + 41 61 306 12 34
E-Mail karger@karger.ch
www.karger.com

Contents Vol. 59, 2002

Brain, Behavior and Evolution

No. 1–2

No. 3

KARGER

Fax + 41 61 306 12 34
E-Mail karger@karger.ch
www.karger.com

Access to full text and tables of contents, including tentative ones for forthcoming issues:
www.karger.com/journals/bbe/bbe_bk.htm

No. 4

Multimodal Sensory Guidance of Complex Behaviors
Symposium held at the 6th International Congress of Neuroethology, Bonn, July 29–August 3, 2001
Guest Editors: Coombs, S.; New, J.G. (Chicago, Ill.)

No. 5–6

Evolutionary Convergence as a Tool in Neuroscience
13th Annual Karger Workshop, San Diego, Calif., November 9, 2001
Guest Editors: Eisthen, H.L. (East Lansing, Mich.); Nishikawa, K.C. (Flagstaff, Ariz.)

Brain, Behavior and Evolution

Brain Behav Evol
ISSN 0006–8977
59(5–6) 231–338 (2002)

KARGER